Version 2 (History and Archaeology) of Essentials of Statistical Methods

Version 2 (History and Archaeology) of Essentials of Statistical Methods

T. P. Hutchinson
School of Mathematics and Statistics, University of Sydney

Rumsby Scientific Publishing
P.O. Box 76, Rundle Mall, Adelaide, South Australia 5000

First published September 1993.
ISBN 0 646 15653 5
Rumsby Scientific Publishing, P.O. Box 76, Rundle Mall, Adelaide, South Australia 5000.
(Please order this book from Rumsby Scientific Publishing, P.O. Box Q355, Queen Victoria Building, Sydney, N.S.W. 2000, Australia.)

To the reader

Every author hopes to write clearly and accurately. I certainly do. But what I have specifically aimed for is a volume light enough to be carried around, and cheap enough for every student to afford.

I am not intending to compete with the many excellent introductory textbooks of statistics that are available. Indeed, I encourage you to buy one of these. But so many of them are 500, 700, even 1000 pages long. This has real disadvantages. They are heavy. They are expensive. And they are wordy.

Being concise means I have not given examples of every possible variation on a problem. This is not a book suitable for teaching yourself with. My assumption is that you are attending a course of clearly-presented lectures, and that you are missing neither any of these nor any of the tutorial classes, assignments, and statistical computing that reinforce them. If that is so, I believe you will find this book a very useful "memory-jogger".

A typical introductory statistics course gets as far as some techniques of *inference* — the testing of hypotheses and the construction of confidence intervals. That is the subject of Part III of this book. In preparation for this, the student needs to know about *data description* and about *probability*. These are covered in Parts I and II.

I apologise for any mistakes or misprints.

Version 2 (History and Archaeology)

The original version of this book was called *Essentials of Statistical Methods, in 41 Pages* (ESM41P). The present *Version* differs in two main respects. Firstly, many of the examples are taken from history or archaeology. (I should add that very often the numbers or the calculations that are performed with them have been changed from what is in the research paper referenced.) Other examples, either taken from everyday life or in which the specific context is relevant, have been left as they were in ESM41P. Secondly, there are 152 A5 pages rather than 41 A4 pages. The numbering of the paragraphs, footnotes, equations, and Tables is unchanged from ESM41P; where I have needed to include additional paragraphs, they are identified with the letter x (for extra) — e.g., paragraph 68x comes between paragraphs 68 and 69.

Some readers will wish that I had gone further in adapting the selection and arrangement of the material to the audience I have in mind — those who study the past. But I point out that (a) I would have lost interchangeability with ESM41P, and (b) this *Version* is perfectly usable with a general-purpose statistics course, and may stimulate some students to get interested in the past who would not otherwise have done so.

T.P.H.

This book is structured as shown below. A detailed list of contents begins opposite.

Contents

Proofs

This book emphasises the methods of applied statistics, rather than the proofs of mathematical statistics. But there are some algebraic demonstrations herein, and the list below may be useful. (fn = footnote.)

Part I. Data description

1. **How can we summarise small amounts of data?** In the National Gallery of Art, Washington, D.C., there is a remarkable collection of copper-alloy portrait medals that date from the Italian Renaissance.

> Studying their metallic composition is of interest because of what it may reveal about the state of metallurgy at various dates in the 15th and 16th centuries, because the presence of trace elements may suggest ore sources and trade routes, because original medals may differ from later copies, and so on.
>
> This example is based upon Glinsman and Hayek (1993).

The percentages of tin in 15 portrait medals by Pisanello were found to be: 3.9, 1.8, 1.3, 6.0, 6.5, 2.7, 4.0, 2.4, 1.3, 2.6, 4.3, 2.5, 2.7, 2.6, 2.6. How can we summarise these numbers?

2. Before answering that, let us think a little about how the data was collected.

- Often, the items (medals) in the data are not interesting in themselves, but because they are representative of a broader population. If our conclusions are to apply beyond the particular items we have collected, our data needs to be a *random sample*: every medal had the same chance of being included in the 15 whose compositions were measured, and every possible set of 15 medals had the same chance of constituting our sample.

> The nature of the sample is a matter of particular concern to the historian and the archaeologist, who so often have

to deal with material that happens to be available rather than that obtained by random sampling. In the present instance, the 15 medals were by a particular artist, and held in a particular collection. It might be valid to extrapolate conclusions, or some of them, to other medals by this artist, but it probably would not be valid to claim they also apply to medals by other artists.

For a mixture of good and bad reasons, practical problems of sampling are usually not seriously faced up to in introductory statistics courses. However, it is worth mentioning that the textbook by Freedman et al (1991) is particularly strong on problems of sampling, and takes many of its examples from modern social history.

- How accurately were the measurements made? In this case, we can presume the measurements were to the nearest 0.1 of a percent. (With any new data, it is worth asking oneself whether the measurements were rounded to the *nearest* unit being used, or whether there is any reason to think they may have all been rounded *down*, or all been rounded *up*.)

- Notice particularly that the data is *measurements*, not *counts*. Measurements could be made more accurately (e.g., 3.92 per cent), but counts must always be whole numbers (e.g., 7 medals in a collection, it is not possible to have 7.4 or 6.8).

3. The most important thing about numbers in a list like that here is how big a typical one is. The second most important thing is how variable the numbers are.

4. **Measures of location: the mean.** (Measures of "location" refers to where the numbers are located among all possible numbers.) We calculate the mean by adding the numbers together and dividing by how many numbers there are. For the 15 medals,

Mean percentage tin content

$$= \frac{3.9+1.8+1.3+6.0+6.5+2.7+4.0+2.4+1.3+2.6+4.3+2.5+2.7+2.6+2.6}{15}$$

$$= \frac{47.2}{15}$$

$$= 3.15.$$

If we are referring to our observations as x's, the usual symbol for the sample mean is $\bar{x}$. The formula for the calculation we have just done is

$$\bar{x} = \frac{\Sigma\, x}{n}. \tag{1}$$

We are using x to represent percentage tin content, and n to be the number of medals. Σ is the symbol "sigma", and means "add up". (Here and elsewhere in this book, I will omit the label on x. I hope readers appreciate that when written out more fully, the formula is $\bar{x} = \frac{\Sigma_{i=1}^{n} x_i}{n}$.)

$$\text{The mean} = \frac{\text{Add up all the numbers}}{\text{Number of numbers in the sample}}.$$

Arithmetic mean, and *average*, are other names for this.

[5.] **Measures of location: the median.** Arrange the numbers in order of size. The median is then the middle one.

1.3 1.3 1.8 2.4 2.5 2.6 2.6 2.6 2.7 2.7 3.9 4.0 4.3 6.0 6.5
(8th)

Median percentage tin content = 2.6.

(If there had been an even number of observations, we would have taken the average of the two middle ones.) There are usually as many observations smaller than the median as there are observations that are bigger.

Why I have to say "usually" is that this rule is broken in situations like this:

3 3 3 3 4 4 4 4 5

Middle observation = median = 4, but only one (5) is bigger whereas four (all 3's) are smaller.

By the way, when two or more observations are the same, they are said to be "tied".

6. **Measures of location: the mode.** The mode is the observation that occurs most frequently. For the 15 medals, it is 2.6, which occurred three times. The mode is not usually very useful for measurement data, because most measurements occur only once, and getting three 2.6's was only because we measured to the nearest 0.1. It is more useful for counted data (e.g., number of persons in a household), or as the *modal group* when measurement data has been grouped into ranges. (The modal group is the group for which $\frac{\text{Number of observations in the group}}{\text{Width of the group}}$ is biggest.)

7. **Features of the mean.**

- Easy to calculate.
- Much statistical theory is based upon it (as we shall see in paragraph 136).
- Accurate, in the sense that the means of different samples do not vary very much. (See paragraph 131 for more on this.)
- In a common sense kind of way, each observation contributes equally to the calculation.

But is the last point a good feature or a bad feature?

Suppose the data, actually 3.9 1.8 1.3 6.0 6.5 2.7 4.0 2.4 1.3 2.6 4.3 2.5 2.7 2.6 2.6, were punched into your calculator wrongly, as $\frac{39+1.8+1.3+6.0+6.5+2.7+4.0+2.4+1.3+2.6+4.3+2.5+2.7+2.6+2.6}{15}$

$$= \frac{82.3}{15}$$
$$= 5.49.$$

We can see the mean is sensitive to mistakes.

8. A further point is that we might sometimes suspect people will be interested in the sum of some number of x's — in which case, it will be the mean they will need in order to calculate it.

Suppose we have observed the numbers of occupants in cars on roads leading to the city centre. If we think someone is going to want to calculate the number of persons entering the city centre by car by multiplying the number of cars by the number of people per car,[1] then it is the *mean* they will need, not the median or the mode.

9. Features of the median.

- It can be calculated even when the data is merely graded, not measured.

  For example, we might be interested in the military rank of soldiers participating in a rebellion. We can order the soldiers from the privates to the generals, and find that the median was perhaps sergeant, but we cannot calculate a mean.

- Very easy to calculate for small samples.
- But for moderate and large samples, it is tedious to place all the observations in order of size.
- All the observations have to be known individually before the median can be computed — i.e., a calculator needs to store them all. (Notice that in computing the mean, the calculator only needs to store the running total $\Sigma\, x$ and n, not all the individual observations.)
- It is not as sensitive as the mean is to mistakes.

1.3 1.3 1.8 2.4 2.5 2.6 2.6 2.6 2.7 2.7 4.0 4.3 6.0 6.5 39

Median = 2.6. In this example, the median was not changed at all by mispunching 3.9 as 39.

---

[1]They might want to use this method because the flow of vehicles along a road is easy to measure automatically, but the number of occupants requires observation by a person.

- Similarly, it is less likely to be disturbed by an observation that is unusually small or large. That is, it accords more closely with our intuition about a typical value than the mean does. As an example, residential house prices are often summarised by the median, since there are a few houses in any city that are 10 times as expensive as most are.

10. **Other measures of location.** These fall into three main classes.

- Modifications to the mean to make it less sensitive to mistakes. For example, discard the biggest 5 per cent of observations and the smallest 5 per cent, then calculate the mean of the others.

- Use the concept of mean on some transformation of the original numbers. Three examples:

  Root mean square $= \sqrt{\text{mean of the squared observations}}$,

  Geometric mean of n positive numbers = nth root of their product,

  Harmonic mean $= \dfrac{1}{\text{mean of the reciprocals of the observations}}$.

  If our dataset consists of just three numbers, 1 2 4,

  $$\text{Mean} = \frac{1+2+4}{3} = 2.33,$$
  $$\text{r.m.s.} = \sqrt{\frac{1+4+16}{3}} = \sqrt{7} = 2.65,$$
  $$\text{g.m.} = (1 \times 2 \times 4)^{1/3} = 8^{1/3} = 2.00,$$
  $$\text{h.m.} = \frac{1}{(1+\frac{1}{2}+\frac{1}{4})/3} = \frac{1}{0.583} = 1.71.$$

  All three of these new "means" obey the equation $T(\text{New mean}) = \text{Mean of } T(x)$, where T is some transformation (square, logarithm, and reciprocal in the three cases).

- Weighted means. Here, each observation is not counted equally, but is "weighted" according to what size it is. Thus we have a weight function $w(x)$, and the weighted mean is $\frac{\Sigma\, w(x).x}{\Sigma\, w(x)}$. The most common example of this is where the observations are weighted according to their sizes, x. The size-weighted mean is thus $\Sigma\, x^2/\Sigma\, x$. Two examples of this may be given.

- Suppose that a third of classes in a university have 9 students in them, a third have 60 students in them, and a third have 300 students in them. So far as a university administrator is concerned, the natural way of calculating the mean class size is as $(9+60+300)/3 = 123$. However, so far as the students are concerned, 9 of them are experiencing a class size of 9, 60 are experiencing a class size of 60, and 300 are experiencing a class size of 300, and the natural way for them to calculate the mean class size is as $(9\times 9+60\times 60+300\times 300)/(9+60+300) = 254$, the size-weighted mean.
- If the time gaps between successive buses are the x's, and people are coming to a bus stop at a rate of λ per unit time, then the λx people arriving during a gap of length x experience an average wait of $\frac{1}{2}x$. So the overall average waiting time for a bus is $\Sigma\,(\lambda x.\frac{1}{2}x)/\Sigma\,\lambda x = \frac{1}{2}\Sigma\, x^2/\Sigma\, x$. (Paragraph 116 has more on this.)

11. **Measures of variation: the range.** The range of a set of numbers is the largest value minus the smallest.

The range of the percentage tin contents is $6.5 - 1.3 = 5.2$.

It is not used much, for the following reasons.

- Only two observations contribute directly to it.
- It is very sensitive to unusually big or small observations.
- There is no standardisation by the size of sample. A bigger sample is likely to have a bigger range than a smaller sample (just because there is a greater chance of including an exceptionally big or exceptionally small observation in the bigger sample).

However, it is sometimes used when simplicity is important and the sample size is always the same. Industrial quality control is a traditional area.[2]

12. Measures of variation: the IQR and the SIQR.

I.Q.R. — inter-quartile range.

S.I.Q.R. — semi-inter-quartile range = $\frac{1}{2}$ IQR.

As its name suggests, the IQR is a range — i.e., a distance between. It is the distance between the *quartiles.*

> The quartiles separate the ordered observations into four equal groups, like the median separates the ordered observations into two equal groups: a quarter of the observations are less than the lower, or first, quartile (LQ), and three-quarters are bigger; three-quarters are less than the upper, or third, quartile (UQ), and one quarter are bigger. The second quartile is the median. (When there are tied observations, some slight changes are needed to this wording.)

It is easy to see where the quartiles are when the number of observations is divisible by 4. But what should we do when this is not the case? Different books say different things. I recommend the following procedure.

> After ordering the observations, the lower quartile is observation no. $0.25n + 0.5$ (where n is the number of observations in the sample), and the upper quartile is observation no. $0.75n + 0.5$.

Thus if there were 10 observations, the lower quartile would be the 3rd and the upper quartile would be the 8th (that is, the 3rd counting from the top). If there were 11 observations, the lower quartile would be observation no. 3.25 (that is, go a quarter of the way from

---

[2] For example, every hour we randomly select n (typically 5) widgets coming off the production line, and check that their measurements are close enough to the specification: firstly, we check their *average*, and secondly, we check that the widgets are not too variable by calculating the *range* of the measurements.

observation no. 3 towards observation no. 4), and the upper quartile would be observation no. 8.75.

13. For these 15 observations,

1.3 1.3 1.8 2.4 2.5 2.6 2.6 2.6 2.7 2.7 3.9 4.0 4.3 6.0 6.5,

the lower quartile is 2.425, the upper quartile is 3.975, so the interquartile range is 3.975 – 2.425 = 1.55.

14. **Measures of variation: the mean absolute deviation (MAD).** The MAD is the mean of the absolute deviations of the observations from the mean.

Observation	Deviation from mean	Absolute deviation from mean
x	$x - \bar{x}$	$\lvert x - \bar{x} \rvert$
3.9	0.75	0.75
1.8	−1.35	1.35
1.3	−1.85	1.85
⋮	⋮	⋮
2.6	−0.55	0.55
	zero	17.95

For this example, the MAD $= \frac{17.95}{15} = 1.20$. (Notice also that the sum of the deviations from the mean is always zero: exactly zero, if the calculations of the mean, the deviations from it, and their mean, are carried to sufficient places of decimals.) In symbols,

$$\text{M.A.D.} = \frac{\Sigma \lvert x - \bar{x} \rvert}{n} \qquad (2)$$

In principle, the MAD is a good measure of variation. But it is not often used, because it is awkward to compute and because mathematical theory to go with it has not been developed.

15. **Measures of variation: the standard deviation.** This is almost (but not quite) the root mean square of the deviations. Here, the above data is repeated, with a column of squared deviations also shown.

Observation x	Deviation from mean $x - \bar{x}$	Squared deviation from mean $(x - \bar{x})^2$
3.9	0.75	0.5625
1.8	−1.35	1.8225
1.3	−1.85	3.4225
⋮	⋮	⋮
2.6	−0.55	0.3025
	zero	32.9173

(Strictly speaking, zero and 32.9173 are the respective sums when the deviations are taken from the exact mean of 3.146..., rather than from 3.15.) For this example, the s.d. = $\sqrt{\frac{32.9173}{14}}$ = 1.53. In symbols,

$$s = \sqrt{\frac{\Sigma\,(x - \bar{x})^2}{n - 1}}. \qquad (3)$$

Notice that we divide the sum of squared deviations by $n - 1$, not by n. (Actually, some books do divide by n, and many calculators have both s_{n-1} and s_n buttons; for more on this, see paragraph 142.)

16. The s.d. is what is most commonly used as a measure of variation. The square of the s.d. has a special name — it is called the *variance*.[3]

17. The above formula (3) demonstrates the meaning of the s.d. — it is based upon the deviations from the mean. But it is rather inconvenient for actually carrying out the calculation: it is necessary to go through the data once to calculate the mean, and then go through it again to calculate the deviations from the mean, and hence the s.d. Furthermore, even if the original dataset consists of whole numbers, the mean will not usually be a whole number, so

[3]We have by now met a number of words that have specialist meanings in statistics — e.g., mean, median, mode, range, standard deviation, variance, quartile. It is a worthwhile exercise to look up such words in an ordinary dictionary, and think how you could improve the definition you find there.

the deviations from it are not whole numbers, and the process of calculating $\Sigma\,(x - \bar{x})^2$ gets messy and error-prone. Instead of the above formula, therefore, the following one is often used.

$$s = \sqrt{\frac{\Sigma\,x^2 - \frac{1}{n}\,(\Sigma\,x)^2}{n-1}} \qquad (4)$$

It gives exactly the same answer, as may be demonstrated with the above data.

x	x^2
3.9	15.21
1.8	3.24
1.3	1.69
⋮	⋮
2.6	6.76
47.2	181.44

From the above table,

$$\begin{aligned} s &= \sqrt{\frac{181.44 - \frac{1}{15}(47.2)^2}{14}} \\ &= \sqrt{\frac{32.9173}{14}}, \end{aligned}$$

which is 1.53, as before.[4]

[4]An algebraic demonstration of the equivalence of the two methods runs as follows. When expanded, $(x - \bar{x})^2$ is $x^2 - 2\bar{x}x + \bar{x}^2$. Summing, we find

$$\begin{aligned} \Sigma\,(x - \bar{x})^2 &= \Sigma\,x^2 - 2\bar{x}\Sigma\,x + n\bar{x}^2 \\ &= \Sigma\,x^2 - \tfrac{2}{n}(\Sigma\,x)^2 + n\frac{(\Sigma\,x)^2}{n^2} \\ &= \Sigma\,x^2 - \tfrac{1}{n}(\Sigma\,x)^2, \end{aligned}$$

as required.

18. However, it is necessary to be careful when using this formula. When calculating $\Sigma x^2 - \frac{1}{n}(\Sigma x)^2$, one number is being subtracted from another. It may happen that the second is only a little smaller than the first. So both need to be worked out quite accurately in order that the difference be sufficiently accurate. Incidentally, Σx^2 is always greater than or equal to $\frac{1}{n}(\Sigma x)^2$. If in your computations this is not so — so that in calculating s you are trying to take the square root of a negative number — you have made a mistake.

19. Notice roughly what size the s.d. is, compared with the scatter in the data. Typically,

> We find a few observations further than 1 s.d. from the mean in a small dataset (e.g., 10 observations);
>
> We find a few further than 2 s.d.'s from the mean in a medium-sized dataset (e.g., 100 observations);
>
> We find a few further than 3 s.d.'s from the mean in a large dataset (e.g., 1000 observations).

20. **Pictorial presentation: the box-and-whisker plot.** In its simple form, this consists of a scale, with beside it a box extending from the lower quartile to the upper quartile, with the median shown, and with whiskers reaching as far as the lowest and highest values. Some people draw it horizontally:

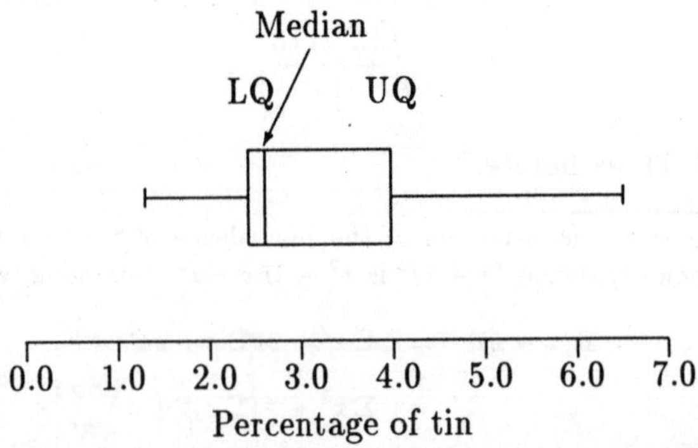

whilst others prefer vertically:

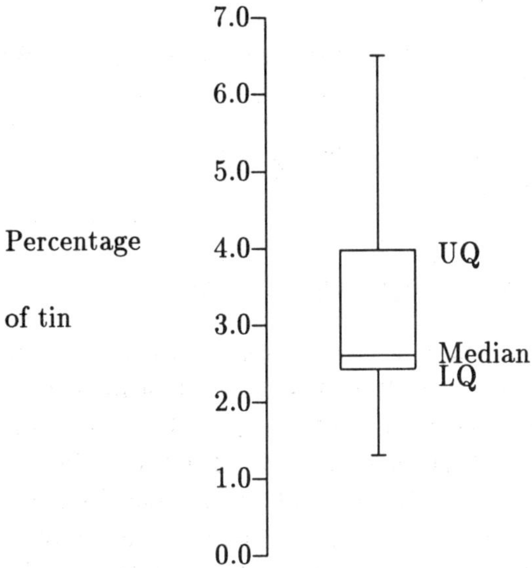

The box-and-whisker plot brings home to us the fact that, in this data, the median is much closer to the lower quartile than to the upper quartile. This is a little unusual, and it makes us look back at the original numbers. We see there are 7 bunched closely together in the middle (2.4, 2.5, 2.6, 2.6, 2.6, 2.7, 2.7), with 3 below and 5 above. This makes us wonder if there is something special about the group of 7 medals. Are they also similar as regards their percentages of zinc, lead, and trace elements? Were they made using the same metallurgical process? Or for a purpose distinct from that of the other 8? Or at about the same date?

21. **Pictorial presentation: box-and-whisker plots that show outliers.** Consider the following dataset of 10 numbers:

8   24   36   40   43   60   68   79   117   127.

There is a big gap between the 8th and 9th numbers, 79 and 117. Perhaps the two largest values, 117 and 127, should not be considered typical of the data as a whole, but should be considered to be *outliers*?

22. A possible definition of an outlier is an observation that is either

> More than 1 IQR above the UQ,

or

> More than 1 IQR below the LQ.

For the above data, we find median = 51.5, LQ = 36, and UQ = 79. Consequently, the IQR is 79 − 36 = 43. Adding this on to 79, we get 122; subtracting it from 36, we get −7. There are no observations below −7, but there is one above 122. Observations which are outliers are now shown on the box-and-whisker plot by means of circles, and the whiskers now extend up as far as the largest observation which is not an outlier (117 in this example), and down as far as the smallest observation that is not an outlier (8 in this example).

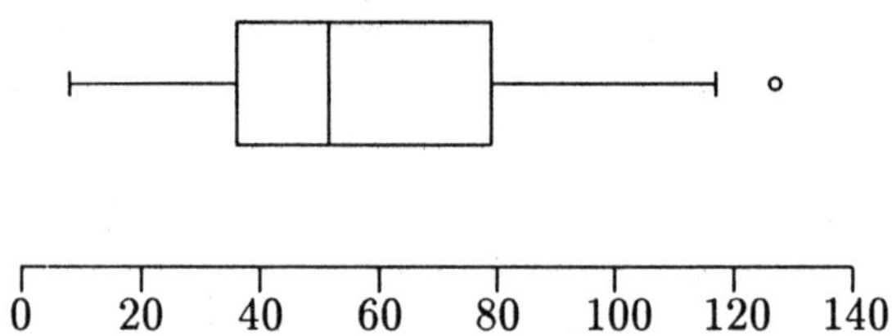

It is notable that here, the division between outliers and the main body of data did not occur where we thought it would: 127, but not 117, was classed as an outlier.

23. **Discussion of outliers.** The above definition of what constitutes an outlier is somewhat arbitrary.

- Even accepting this type of definition, some books use 1.5 IQR's as the step we must go above the UQ or below the LQ before we start to think the observations are outliers.

- The definition takes no account of sample size: in a large sample, one would expect to find some observations more than 1 IQR above the UQ or below the LQ, even if there is nothing genuinely unusual about them.

- Furthermore, this type of definition takes no account of our assessment of how likely it is that a "rogue" observation might arise.

The problem of finding an appropriate definition of an outlier, and of deciding whether a suspect observation is a mistake or not, is a very difficult one, certainly much more advanced than this book. Nevertheless, there are a number of common sense points that are well worth making. When we find an observation that is substantially different from most of the others,

- We should attempt to trace its progress through the procedure of recording the information and transmitting it to us, if it is practicable to do this, as we may find an error has crept in at some stage;

- We should look at the circumstances surrounding the observation, as there may be something of real scientific interest to be discovered;

- We may wish to discard the observation because it is obviously wrong;

- We may wish to accept the observation because we can find nothing amiss with it and it is not too different from the others;

- We may choose to use robust statistical procedures — i.e., ones that will not be greatly harmed if some observations are indeed rogue values. For example, we may prefer the median to the mean, and the IQR to the s.d.

24. **Transformations.** It is usually easiest to understand data oneself, and communicate it to others, if in some sense it is roughly equally spaced out over its range of variation, not all bunched up at one end. So if this is not true of the raw data, we look for some sort of transformation (e.g., taking the square root, or taking the logarithm) that achieves it.[5]

[5] An additional reason for carrying out a transformation is the following. It sometimes happens that we have several different groups of observations in our

25. The following numbers are the percentages of lead in 14 sickles from a Late Bronze Age hoard found in Slovenia (Trampuž-Orel et al, 1991).

0.19 0.33 0.27 0.22 0.41 0.52 0.52 0.06 0.84 0.25 0.40 0.21 0.15 0.54.

In order of size, these are

0.06 0.15 0.19 0.21 0.22 0.25 0.27 0.33 0.40 0.41 0.52 0.52 0.54 0.84.

So the median is 0.30, the LQ is 0.21, and the UQ is 0.52. The IQR is therefore 0.31, and we flag any observations over 0.83 or below −0.10 as being possible outliers. The box-and-whisker plot looks like this:

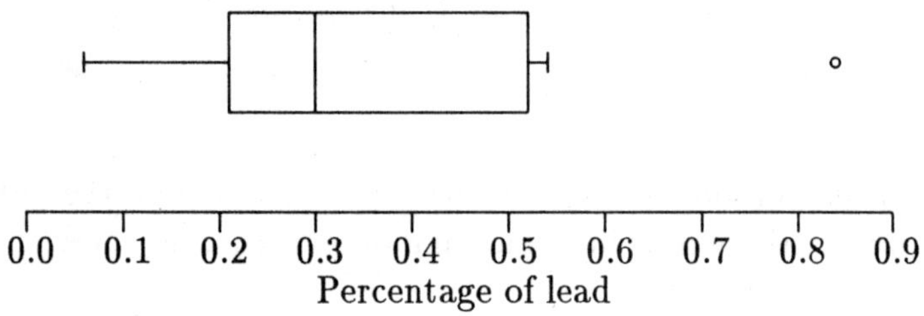

Taking the logarithm of each of the observations, we get

−1.22 −0.82 −0.72 −0.68 −0.66 −0.60 −0.57 −0.48 −0.40 −0.39 −0.28 −0.28 −0.27 −0.08

The median is −0.525, the LQ is −0.68, and the UQ is −0.28. So the IQR is 0.40, and there is now an outlier at the lower end, but not at the upper! The box-and-whisker plot looks like this:

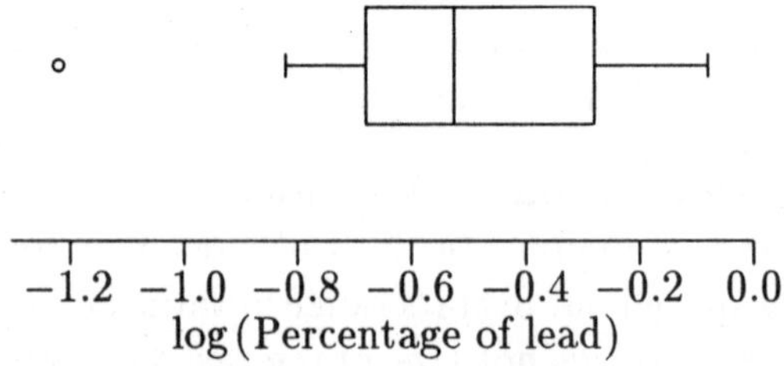

dataset, and these have different standard deviations; some statistical procedures (beyond the scope of this book) require the assumption of equal standard deviations (homoscedasticity). It is sometimes possible to find a transformation such that the s.d.'s in the groups are approximately equal after the transformation has been made.

A disadvantage of presenting the data in transformed form is that the numbers themselves are now no longer so meaningful. This can be remedied by showing a scale of the original numbers:

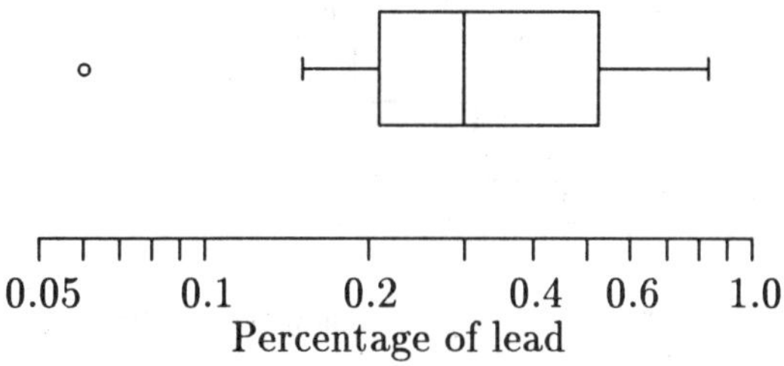

(0.05 is where $-1.301 = \log(0.05)$ was on the previous graph, 0.1 is where $-1.0 = \log(0.1)$ was, and so on.) If the square root transformation is used instead, there are no outliers at all. In choosing whether to make a transformation, and which one to choose, there is usually no "right" or "wrong" answer. It is not even necessary to choose one that leaves no outliers showing — after all, one or more of the observations may truly be an error. Usually, we search for a transformation that (a) is mathematically simple (e.g., the logarithm and the square root), and/or has some physical meaning in the context of the problem being studied, and (b) makes the appearance of the box-and-whisker plot fairly symmetric.

26. **Skewness.** The skewness of a set of numbers is how far from symmetric they are. There are various ways of defining it more precisely. One that is easy and directly reflects what is shown in the box-and-whisker plot is the ratio $\frac{\text{Upper quartile} - \text{Median}}{\text{Median} - \text{Lower quartile}}$. Evidently, a ratio of 1 corresponds to symmetry. For the original data, this is $\frac{0.52-0.30}{0.30-0.21}$, which equals 2.44; for the logarithmically-transformed data, it is rather less, $\frac{-0.28-(-0.525)}{-0.525-(-0.68)} = 1.58$.

27. **Pictorial presentation: the cumulative frequency plot.** In a cumulative frequency plot, the vertical axis represents the number (or proportion, percentage, etc.) of observations that are less than or equal to the value x shown on the horizontal axis. For the percentages of lead in sickles, the vertical axis in the graph to be shown now has been marked with the *proportion* of sickles whose lead content is less than or equal to x, but it could instead be marked

with the *percentage*, or with the *number*. (As a reminder, the data has been listed at the lower right side of the graph.)

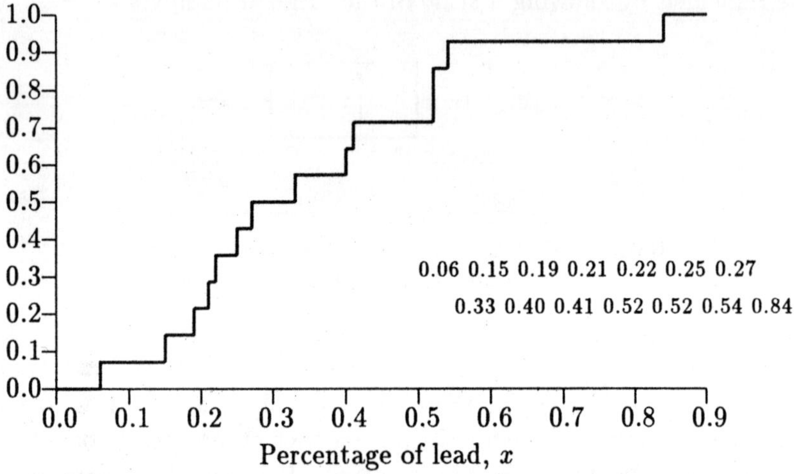

28. If we do not want to show it as a series of steps, a reasonable alternative is to use the midpoints of the verticals as being the "plotting positions", and join these up with straight lines. That is, we are plotting the ith smallest observation, x, at the point $\left(x, \frac{i-\frac{1}{2}}{n}\right)$. In the following illustration, the steps have been left on the graph, to illustrate what is being done.

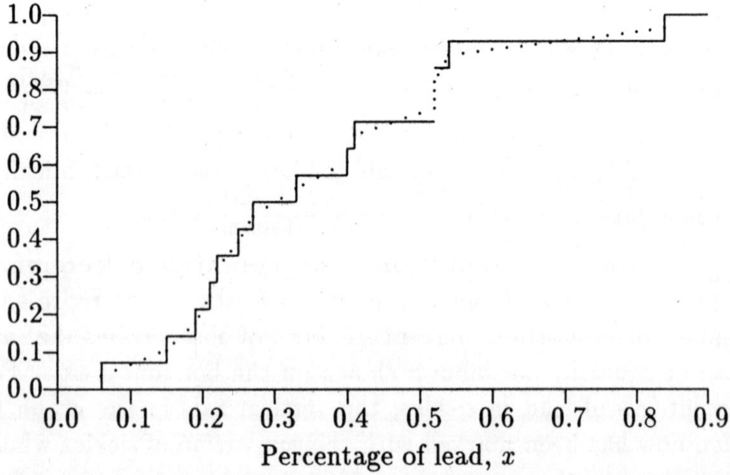

Notice that when we have done this, the graph no longer represents exactly what is true of the sample of data we have. Instead, it is an estimate of what is true of the population from which the sample was taken. As to how to compare data plotted like this with a theoretical distribution, such as the normal, see paragraph 107 below.

29. With the above choice of plotting positions, the quartiles will be the same as defined above in paragraph 12. For example, the lower quartile is the value of x for which the cumulative proportion reaches 0.25. So, on setting $0.25 = \frac{i-\frac{1}{2}}{n}$, we find $i = 0.25n + 0.5$, as in paragraph 12.[6]

30. **Why no histograms yet?** You may have heard about histograms, and wondered why we have not drawn any yet. The reason is that an appreciable number of observations is needed before a respectable histogram can be drawn. See paragraph 35 below for histograms.

31. **Summarising moderate amounts of data.** So far, we have had so little data that we have been able to consider the observations one by one. Now, what if there are a few tens or hundreds of observations, how do we summarise them?

> If we have many thousands or millions of observations, we get a computer to do the classifying and counting and tabulating for us. Processing data like this is a subject that is practically important, but is beyond the scope of this book.

32. **Tallying and tabulating.** We shall see in paragraph 105x that the size and shape of humans provides evidence relevant to some interesting historical questions. (If people were big, they were probably well-fed, and hence probably economically prosperous.) Today, the size and shape of humans is important for, among other things, the design of the interior of motor vehicles. In an anthropometric survey of 100 male car drivers, the following arm length measurements (mm.) were found.

---

[6] Not every book defines the quartiles in exactly the way given in paragraph 12, and the plotting positions in exactly the way given here, and in some books the quartiles obtained from the direct definition work out to be slightly different from the ones obtained by seeing when the cumulative proportion graph reaches 0.25 and 0.75.

742	817	846	845	846	833	782	767	786	810
765	694	758	754	754	806	775	798	740	809
759	785	795	830	854	830	789	802	720	816
764	783	747	774	763	781	804	727	809	801
796	791	811	833	757	786	806	796	776	803
801	817	831	811	801	802	834	805	829	817
801	769	706	802	774	767	811	767	830	771
759	751	765	811	727	761	808	777	835	787
788	776	754	812	860	765	763	780	777	737
761	791	757	758	795	708	784	725	800	723

To tabulate this data, we first choose suitable classes, and then proceed systematically through the numbers, placing a tally mark in the appropriate class and crossing off the number when we have done so. (This is because someone always comes in and interrupts us, and we would otherwise forget where we had got to!)

up to 699	/
700–719	//
720–739	~~////~~ /
740–759	~~////~~ ~~////~~ ///
760–769	~~////~~ ~~////~~ //
770–779	~~////~~ ///
780–789	~~////~~ ~~////~~ /
790–799	~~////~~ //
800–809	~~////~~ ~~////~~ ~~////~~ /
810–829	~~////~~ ~~////~~ /
830–849	~~////~~ ~~////~~ /
850–869	//
870 & over	

And we count up the tally marks to get the frequencies:

up to 699	1
700–719	2
720–739	6
740–759	13
760–769	12
770–779	8
780–789	11
790–799	7
800–809	16
810–829	11
830–849	11
850–869	2
870 & over	0

33. In choosing the classes, we need to bear in mind the following points:

- Usually, convenient round numbers are used.
- Sometimes, one or more specific classes must be identifiable in the tabulation (e.g., information about them may be required for some definite reason, they may need to be compatible with other statistical tabulations, or simply be what is customary). For example, when tabulating the ages of road casualties, the age at which a motor vehicle can legally be driven and the age at which alcohol can legally be consumed are both of particular interest.
- There need to be sufficiently many classes that we do not lose too much information because of crude, wide, grouping. But there must also be sufficiently few classes that we are able to comprehend the message of the data. Between 5 and 15 classes will usually be suitable.
- Each observation must be able to go into a class. (If the measurements had been recorded to one decimal place of accuracy, we would have to use one decimal place of accuracy in our list of classes — with the above classes, we would not know where to put observations of 699.7 or 719.4, for instance.)
- Each observation must go into only one class. (If the classes had been up to 700, 700–720, 720–740, etc., we would not know where to put observations of 700 or 720.)
- Ideally, we should know exactly what range of exact measurements each class represents. (Thus, do the above classes correspond to exact arm lengths of up to 700, 700–720, 720–740, etc., the measurements having been rounded down? Or, do they correspond to exact arm lengths of up to 699.5, 699.5–719.5, 719.5–739.5, etc., the measurements having been rounded to the nearest mm.?)

  Why I say "ideally" is because all too often, this information is lost to us. But this does not generally matter too much, because rounding error is usually a tiny fraction of all the other errors and variability that might be present.

34. **Stem-and-leaf plots.** The stem-and-leaf plot has the advantage, as compared with an ordinary tally, of retaining information about the final digit. (Hence any over-representation of the digits 0 and 5 should be noticeable, especially if a second version of the plot is prepared, with the final digits sorted into order.) It is easier to show what a stem-and-leaf plot is than to spend many words describing it. For the data in paragraph 32 it looks like this::

```
69 | 4
70 | 6 8
71 |
72 | 0 7 7 5 3
73 | 7
74 | 2 0 7
75 | 8 4 4 9 7 9 1 4 7 8
76 | 7 5 4 3 9 7 7 5 1 5 3 1
77 | 5 4 6 4 1 7 6 7
78 | 2 6 5 9 3 1 6 7 8 0 4
79 | 8 5 6 1 6 1 5
80 | 6 9 2 4 9 1 6 3 1 1 2 5 1 2 8 0
81 | 7 0 6 1 7 1 7 1 1 2
82 | 9
83 | 3 0 0 3 1 4 0 5
84 | 6 5 6
85 | 4
86 | 0
```

35. **Pictorial presentation: histograms.** The table of frequencies given at the end of paragraph 32 is shown as a histogram below:

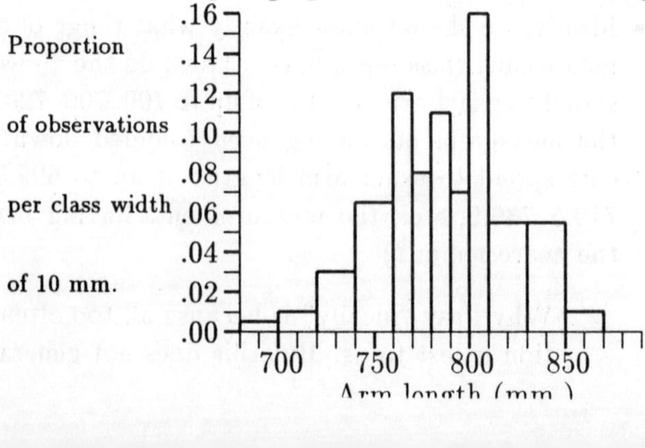

Notice the following points about histograms:

- The x-axis is drawn as a scale, with appropriate values marked on it.

- If the table does not exactly specify what the lowest and highest groups exactly are, it is necessary to make some sensible assumption. I have taken "up to 699" to be 680–699.

- The height of each bar is proportional to the frequency in the class divided by the width of the class. In this example, a class width of 10 mm. has been chosen as the standard; classes like 700–719 and 810–829 are of double width, so their height is half their frequency, whilst the height of classes like 760–769 and 800–809 equals their frequency. If you get this wrong in an examination, the examiner is likely to conclude you have never been to any lectures, never attended any tutorial or practical classes, never opened any textbook, and, in particular, never before attempted to draw a histogram yourself. After all, to draw a histogram in the correct fashion is only a matter of educated common sense. Let us look at the observations marked on a scale of length:

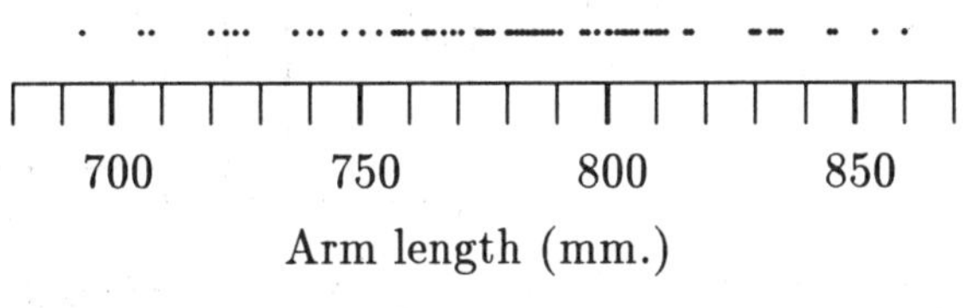

The idea that we are capturing with a histogram is how dense are the observations in any region along the scale — that is, how many observations there are per unit along the scale. So obviously we must standardise on a particular class width.

- Label the y-axis appropriately — to make it clear that it is a proportion of observations (or percentage, or number) *per unit on the x-axis.* (It is common for books to tell you to include a scale of area on the histogram. But I think people looking at a graph do not appreciate a scale of area very well, and I prefer careful labelling of the y-axis.)

- There are no gaps between the bars (between 699 and 700, or between 719 and 720, for example).[7]
- Because the x-axis is marked as a scale, there is no need to label the bars (as up to 699, 700–719, and so on).
- But if the quantity being tabulated is necessarily a whole number, it is sensible to label the bars, rather than marking a scale. Suppose that among 708 bus drivers, 224 of them had 0 accidents in a two-year period, 226 had 1 accident, 150 had 2 accidents, 68 had 3 accidents, 40 had 4–7 accidents, and none had 8 or more accidents. Then I would draw the histogram like this:

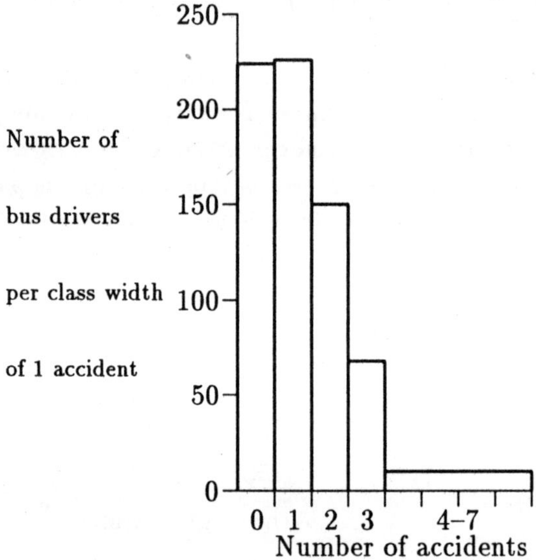

36. **Pictorial presentation: the cumulative frequency plot (again).** The cumulative frequency table for the arm length data is as follows:

---

[7]If you can assume that

up to 699	really means	up to 699.5,
700–719	really means	699.5–719.5,

and so on, some instructors will tell you to put the bases of the bars over the ranges

up to 699.5	instead of	up to 700,
699.5–719.5	instead of	700–720,

and so on. But to my mind, the ease of communicating up to 700, 700–720, etc., to other people outweighs the strict correctness of up to 699.5, 699.5–719.5, etc.

up to 699	1	(.01)
up to 719	3	(.03)
up to 739	9	(.09)
up to 759	22	(.22)
up to 769	34	(.34)
up to 779	42	(.42)
up to 789	53	(.53)
up to 799	60	(.60)
up to 809	76	(.76)
up to 829	87	(.87)
up to 849	98	(.98)
up to 869	100	(1.00)

I have also shown the cumulative *proportions*, in brackets. When plotted, the graph looks like this:

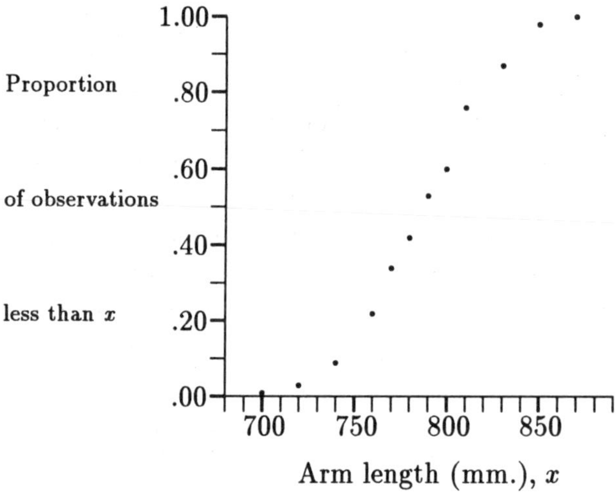

(Some people like to join up the points with straight lines.) When drawing a graph like this, it is always necessary to think about exactly what the y-axis should be labelled — in particular, is it the proportion *less than* x that is being shown, or is it the proportion *less than or equal to* x that is being shown?

> I have plotted the points at arm lengths of 700, 720, etc., because these are round numbers and round numbers aid in communicating the message of data to other people.

Some instructors prefer you to plot them at 699.5, 719.5, etc. The justification is that these are the upper limits of the actual arm lengths in the various classes, the actual arm length having been recorded to the nearest mm.

Other instructors prefer you to plot them at 699, 719, etc., and to label the y-axis with "less than or equal to", rather than with "less than". The justification is that this most closely represents what is true of the recorded measurements.

What you must not do, of course, is to plot the points at the mid-points of the classes (690, 710, etc.) — it is not true, nor is it anywhere near true, that a proportion .03 of observations are less than 710, .09 less than 730, etc.

As to how to compare data plotted like this with a theoretical distribution, such as the normal, see paragraph 107 below.

37. **Computing the mean and standard deviation of grouped data.** If we have a table of frequencies like the arm length data above, how do we calculate the mean and s.d.? The basis of the method is to assume that all the observations in a class were in fact at its mid-point (denoted x below), so that we had 1 observation of 690, 2 of 710, 6 of 730, etc. We set out the calculations as below.[8]

Class	x	Freq., f	xf	x^2f
up to 699	690	1	690	476100
700–719	710	2	1420	1008200
720–739	730	6	4380	3197400
740–759	750	13	9750	7312500
760–769	765	12	9180	7022700
770–779	775	8	6200	4805000
780–789	785	11	8635	6778475
790–799	795	7	5565	4424175
800–809	805	16	12880	10368400
810–829	820	11	9020	7396400
830–849	840	11	9240	7761600
850–869	860	2	1720	1479200
		100	78680	62030150

[8] Some calculators permit x and f to both be input, and so a table of xf and x^2f like this becomes unnecessary.

Notice the following three points:

- The quantities in the final column are $x^2 f$, not $(xf)^2$. That is, they are the quantities in the fourth column multiplied by the respective quantities in the second column; they are *not* the squares of the quantities in the fourth column.

- The x for the "up to 699" class is something of a guess, of course. To take it as 695 or 680, instead of 690, would be reasonable. To take it as 350 (midway between 0 and 700) would be daft.

- Some instructors may want you to take the x's as 689.5, 709.5, etc., for reasons already discussed. The mean which they will calculate will be 0.5 less than the mean which we will get (because they are effectively assuming each observation is 0.5 less than we are); the standard deviation which they will calculate will be identical to the standard deviation which we will get (because reducing each observation by 0.5 makes no change to how variable the observations are).

Naturally, the results of calculations based on grouped data will not be exactly the same as those on raw data.

38. The mean is

$$\frac{1}{n}\Sigma\, x f, \tag{5}$$

which in this example is $\frac{1}{100}78680 = 786.80$. The s.d. is 35.49, which is calculated as follows:

$$s = \sqrt{\frac{\Sigma\,(x-\bar{x})^2 f}{n-1}} \tag{6}$$

$$= \sqrt{\frac{\Sigma\, x^2 f \;-\; \frac{1}{n}(\Sigma\, x f)^2}{n-1}} \tag{7}$$

$$= \sqrt{\frac{62030150 \;-\; \frac{1}{100}(78680)^2}{99}}$$

$$= \sqrt{\frac{124726}{99}}$$

$$= 35.49 \text{ (mm.)}$$

The first formula is the one that best explains what a standard deviation is, the second formula is the one that is most convenient for actually doing the calculation.[9] This is just like the two formulae we had when calculating the s.d. in paragraphs 15 and 17. Indeed, notice that if we know all the observations exactly, rather than them merely being placed into classes, all the f's will be 1, and the above formulae will be the same as the ones we had in paragraphs 15 and 17.

39. This type of calculation is sometimes made easier by subtracting some constant number from all of the x's, and/or by dividing all the x's by some constant number. Try the calculation with 700 subtracted from all the x's. You will get $\bar{x} = 86.80$ (to which 700 needs to be added to get the mean arm length), and s.d. $= 35.49$ (which is the s.d. of arm length, as subtracting the same amount from each observation makes no difference to how variable they are). Computational tricks like this used to be very important, but are less so now, with the ready availability of electronic calculators.

40. I have given the values of mean and s.d. to more decimal places than can be justified: if we took another sample of 100 male car drivers, their mean arm length could be different from 786.80 by 5 or 10 mm. This issue of sampling variability will be taken up from paragraph 128 onwards. Of course, the point of giving unjustifiable decimal places is so that the reader can check the working.

41. **Computing the median and quartiles of grouped data.** If we have joined the points on the cumulative proportion graph with straight lines, we simply see where a cumulative proportion of .50 is reached (in order to read off the median), and where cumulative proportions of .25 and .75 are reached (in order to read off the quartiles). For the above example, we plotted a cumulative proportion of .42 at $x = 780$, and a cumulative proportion of .53 at $x = 790$. A cumulative proportion of .50 is therefore reached at $x = 780 + \frac{.08}{.11}(790 - 780) = 780 + \frac{8}{11}10 = 787.27$. Similarly, cu-

[9] But remember the point made in paragraph 18. When calculating $\Sigma x^2 f - \frac{1}{n}(\Sigma x f)^2$, one number is being subtracted from another. It may happen that the second is only a little smaller than the first. So both need to be worked out quite accurately in order that the difference be sufficiently accurate. Remember also that the second is *always* less than or equal to the first.

mulative proportions of .22 and .34 were plotted at $x = 760$ and $x = 770$ respectively, so a cumulative proportion of .25 is reached at $x = 760 + \frac{.03}{.12}(770 - 760) = 760 + \frac{3}{12}10 = 762.50$. And a cumulative proportion of .75 is reached at $x = 809.37$. The inter-quartile range is therefore $809.37 - 762.5 = 46.87$.

42. **Quintiles, deciles, percentiles.** Just like quartiles divide the ordered observations into four equal groups, quintiles divide the ordered observations into five equal groups, deciles divide the ordered observations into ten equal groups, and percentiles divide the ordered observations into one hundred equal groups. A woman of 5th percentile height is one who is taller than 5 per cent of women (and shorter than 95 per cent). A man of 95th percentile height is one who is taller than 95 per cent of men (and shorter than 5 per cent).[10] The 85th percentile vehicle speed is such that 85 per cent of vehicles are slower and 15 per cent are faster.[11]

43. **Plotting one variable against another: regression and correlation.** Suppose that for each value of x, a value of y corresponds, and we plot one against the other. The table below gives x = the distance of various provinces from Guangzhou (in hundreds of kilometres), and y = a measure of how strongly the price of rice in the province depended on the price in Guangzhou during the 18th century — what we might term the linkage between them. (This example was suggested by the paper by Marks, 1991.)

Province	x	y
Qingyuan	5.4	0.2
Pingle	4.0	0.3
Zhen'an	8.1	0.2
Yulin	10.0	0.3
Wuzhou	2.4	0.5
Gaozhu	5.2	0.6

The scatterplot of this data is as follows:

[10] The crash testing of vehicles is often carried out using dummies built to represent the 5th percentile female and the 95th percentile man, as well as ones built to represent someone of average size.

[11] It is an often-used guide to what speed limit to set on a road.

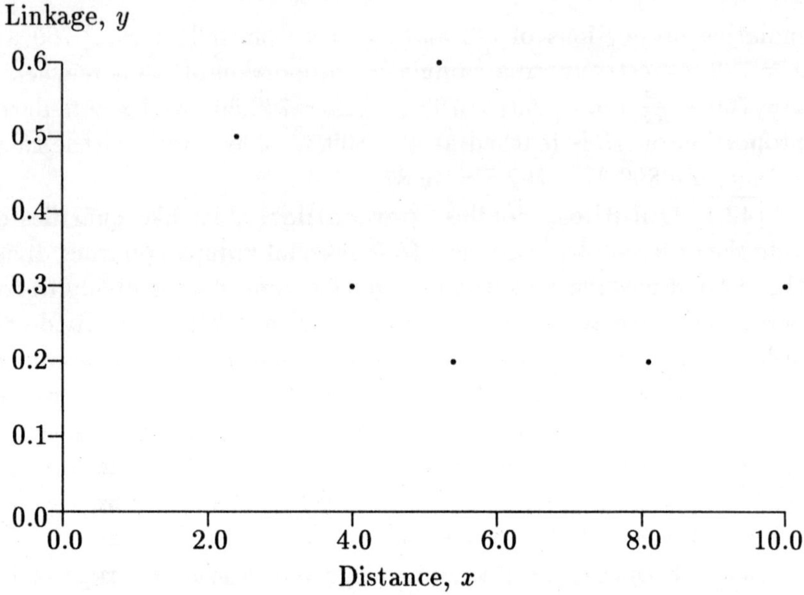

The scatterplot itself is a worthwhile technique of data description. Going a little further, we now study how to predict y from x, and how to measure the strength of the relationship between the variables.

- Regression — with this term, emphasis is on the equation that predicts y from x. We will only be concerned with using a straight line to predict y from x.

- Correlation — with this term, emphasis is on the strength of the relationship. That is, how useful is knowing x for the purpose of predicting y? We will only be concerned with the strength of the *linear* relationship.

44. What is meant by the "best" straight line?

Any straight line obeys the equation $y = a+bx$, for some choice of the constants a and b. So choosing a line means choosing a and b.

The criterion that is most used for deciding whether one line is better than another is the sum of squared errors: that is, the aim is find a and b such that the sum of squared errors is as small as possible. I will demonstrate this using a dataset that is so small that it is easy to do repeated calculations on it.

45. Let x and y be as below:

x	y
1	4
2	2
3	6
4	8

We will consider several alternative prediction lines.

- Firstly, the equation $\hat{y} = 3 + x$ (the hat ˆ is placed on the y as a notation for a predicted value). The table below gives a comparison between the predicted $\hat{y}$ and the observed y, and includes the errors $y - \hat{y}$ and the squared errors, $(y - \hat{y})^2$.

x	y	$\hat{y}$	$y - \hat{y}$	$(y - \hat{y})^2$
1	4	4	0	0
2	2	5	−3	9
3	6	6	0	0
4	8	7	1	1
				10

  So for this line — that is, for the choice of $a = 3$ and $b = 1$ — the sum of squared errors is 10.

- Secondly, $\hat{y} = 1 + 2x$. The predictions and errors are as shown:

x	y	$\hat{y}$	$y - \hat{y}$	$(y - \hat{y})^2$
1	4	3	1	1
2	2	5	−3	9
3	6	7	−1	1
4	8	9	−1	1
				12

  So for this line — that is, for the choice of $a = 1$ and $b = 2$ — the sum of squared errors is 12, a little worse than for our first choice of line.

- Thirdly, $\hat{y} = 8 - x$. The predictions and errors are as shown:

x	y	$\hat{y}$	$y-\hat{y}$	$(y-\hat{y})^2$
1	4	7	−3	9
2	2	6	−4	16
3	6	5	1	1
4	8	4	4	16
				42

So for this line — that is, for the choice of $a = 8$ and $b = -1$ — the sum of squared errors is 42, much worse than for our previous lines.

The above demonstrates what the sum of squared errors means.

46. **Formulae for b and a.** The values of a and b for which the sum of squared errors is as small as possible are obtained as follows.

Calculate the slope b from

$$b = \frac{\Sigma\, xy - \frac{1}{n}\Sigma\, x \Sigma\, y}{\Sigma\, x^2 - \frac{1}{n}(\Sigma\, x)^2}. \tag{8}$$

Then, knowing b, calculate the intercept a from

$$a = \bar{y} - b\bar{x} \tag{9}$$

(where $\bar{y}$ and $\bar{x}$ are the means of y and x, of course).[12]

For the dataset of four points that we have been discussing, the calculations are as follows.

[12] Equations (8) and (9) can be derived by the method below.

- The predicted value is $\hat{y} = a + bx$. So the sum of squared errors $\Sigma\,(y - \hat{y})^2$ is $\Sigma\,(y - a - bx)^2$. We want to find the values of a and b that minimise this.
- To find a minimum, we differentiate and set the resulting equation equal to 0. Of course, we differentiate with respect to a and with respect to b — the x's and y's are known, specific, observations. Differentiating with respect to a, we find that $\Sigma\,(y - a - bx)$ needs to equal 0; differentiating with respect to b, we find that $\Sigma\, x(y - a - bx)$ needs to equal 0.
- We thus have two equations in two unknowns (a and b). On solving these equations by standard methods and rearranging, we obtain the equations for b and a.

x	y	x^2	xy
1	4	1	4
2	2	4	4
3	6	9	18
4	8	16	32
10	20	30	58

Firstly, calculate the slope:

$$\begin{aligned} b &= \frac{58 - \frac{1}{4} \times 10 \times 20}{30 - \frac{1}{4} \times 10 \times 10} \\ &= \frac{8}{5} \\ &= 1.6. \end{aligned}$$

Secondly, the intercept:

$$\begin{aligned} a &= 5 - 1.6 \times 2.5 \\ &= 1. \end{aligned}$$

Thus $\hat{y} = 1 + 1.6x$ is the least squares line. The sum of squared errors turns out to be 7.2:

x	y	$\hat{y}$	$y - \hat{y}$	$(y - \hat{y})^2$
1	4	2.6	1.4	1.96
2	2	4.2	−2.2	4.84
3	6	5.8	0.2	0.04
4	8	7.4	0.6	0.36
				7.20

(The above method for finding the slope and the intercept is one that is sensitive to outliers. There are methods that are less affected by outliers, but they are not in wide use at present.)

47. **Convenient computation of b and a.** Here, the data on linkage between markets for rice is repeated.

x	y	x^2	y^2	xy
5.4	0.2	29.16	0.04	1.08
4.0	0.3	16.00	0.09	1.20
8.1	0.2	65.61	0.04	1.62
10.0	0.3	100.00	0.09	3.00
2.4	0.5	5.76	0.25	1.20
5.2	0.6	27.04	0.36	3.12
35.1	2.1	243.57	0.87	11.22

- Preliminary calculations are as above.

- Calculate S_{xx}, S_{yy}, and S_{xy}, defined as:

$$S_{xx} = \Sigma x^2 - \tfrac{1}{n}(\Sigma x)^2 \qquad (10)$$
$$S_{yy} = \Sigma y^2 - \tfrac{1}{n}(\Sigma y)^2 \qquad (11)$$
$$S_{xy} = \Sigma xy - \tfrac{1}{n}(\Sigma x)(\Sigma y). \qquad (12)$$

For this data,

$$S_{xx} = 243.57 - \tfrac{1}{6} \times 1232.01 = 38.235$$
$$S_{yy} = 0.87 - \tfrac{1}{6} \times 4.41 = 0.135$$
$$S_{xy} = 11.22 - \tfrac{1}{6} \times 35.1 \times 2.1 = -1.065.$$

- Now, $b = S_{xy}/S_{xx} = -1.065/38.235 = -0.0279$, and $a = 0.35 - (-0.0279)5.85 = 0.513$.

- The quantities S_{xx}, S_{yy}, and S_{xy} are used in various other calculations associated with regression lines, also; see paragraphs 51 and 183.

So $\hat{y} = 0.513 - 0.0279x$ is the regression line of y on x. (Looking at the plot of the points in paragraph 43, a negative value of b is obviously reasonable.)

48. **Residuals.** Having found the best-fitting straight line, $\hat{y} = a + bx$, it is sensible to take a look at the pattern of "residuals", $y - \hat{y}$. Suppose we plot these against x.

- What we are usually hoping to see is no pattern — a completely random scatter of points. This means that there is no reason to invent a more complicated model for predicting y from x than the straight line equation.

- Sometimes we see the points lie on a curved line. This means that the dependence of y on x is more complicated than our straight line equation — a quadratic equation $y = a + bx + cx^2$, or other more complicated equation, would be better.[13]

[13] In such a case, the curvature of the points can be seen on the original graph of y plotted against x, too. But it is typically easier to spot it on the graph of $y - \hat{y}$ plotted against x.

- Sometimes we see the amount of scatter is greater for some values of x than for others — for example, that higher values of x are associated with greater scatter. Clearly, it is well worth knowing that x is a poorer predictor for some values than for others.

49. **What if we want to predict x from y?** If we want an equation $\hat{x} = c + dy$ for predicting x from y, we find c and d as follows:

$$d = \frac{\Sigma\, xy - \frac{1}{n}\Sigma\, x \Sigma\, y}{\Sigma\, y^2 - \frac{1}{n}(\Sigma\, y)^2} \tag{13}$$

$$c = \bar{x} - d\bar{y}. \tag{14}$$

For the little dataset of four points that we have been using, $d = 0.4$ and $c = 0.5$. Hence $\hat{x} = 0.5 + 0.4y$ is the best straight line for predicting x from y.

> If we rearrange this to make y the subject, we find $y = -1.25 + 2.5\hat{x}$. This is different to the best equation for predicting y from x. The two regression lines are different.

50. The fact that the regression line of x on y is different from the regression line of y on x is by no means a mere academic curiosity. As an instance of its importance, there was a controversy in the U.K. a few years ago concerning the saturation level of car ownership — that is, the number of cars per person when everybody who wants a car does in fact have one. As I recall it, the controversy was roughly as follows.

- Highways are planned many years in advance of their construction. One of the important factors going into the decision of what length of highways to plan for is the number of cars that people will own 20 or 30 years in the future. One method of estimating this is to look at a graph of y = rate of growth of car ownership, against x = level of car ownership.

- It was found that areas of the U.K. having high levels of car ownership (primarily, the more prosperous areas) mostly had low rates of growth of car ownership; this was interpreted as meaning that nearly everyone there who wanted a car had already bought one. But areas having low levels of car ownership (less prosperous areas) mostly had high rates of growth of car ownership.

- One could therefore find an equation $y = a + bx$ (with b being negative) that had significant predictive power. Official transport planners did this, and used it to determine what value of x (car ownership) would correspond to $y = 0$ (that is, zero growth in car ownership).

- They were then criticised (by people who could be characterised as environmentally conscious, or anti-roads) on the grounds that if x is being found from a known y, then the regression line of x on y should be used. When this was done, the saturation level of car ownership was substantially smaller than before. (In terms of the cost of construction of the highways required, the difference was equivalent to thousands of millions of pounds.)

- The most convincing resolution of the controversy that I saw was one that emphasised that neither of the regressions is exactly what is required; instead, the best description of the data should be sought; and this must take account both of errors in measuring the x's and of errors in measuring the y's.

51. **Correlation.** The correlation coefficient (more fully, the *product-moment* correlation coefficient) measures the strength of linear association between x and y. The formula is

$$r = \frac{S_{xy}}{\sqrt{S_{xx}S_{yy}}}. \tag{15}$$

For the data on linkage of rice markets,

$$r = \frac{-1.065}{\sqrt{38.235 \times 0.135}},$$

which is -0.47.

Regarding the interpretation of the correlation coefficient, the following points are worth making.

- Correlation is always between -1 and 1.

- A correlation of 1 means all the points lie on a perfect straight line, and y increases as x increases; a correlation of -1 means all the points lie on a perfect straight line, and y decreases as x increases.

- A correlation of 0 means that there is no linear relationship. This may be because there is no relationship at all, or it may be there is a relationship that is so strongly curved that there is no linear component to it.

- A positive correlation means that as x increases, so y tends to increase also. A negative correlation means that as x increases, so y tends to decrease.

- The correlation does not change when a constant number is added to all the x's, and/or to all the y's; nor does it change when all the x's and/or all the y's are multiplied by a constant number. But it does change when a *non-linear* transformation is applied to the x's and/or to the y's.

- The correlation does not measure how much y changes with a unit change in x. The slope b is what measures that.

- The finding of a correlation between x and y does not prove anything about *causation.* There are many possibilities. It might indeed be that changes in x are causing changes in y. But it might be that changes in y are causing changes in x. Or there might be some third factor that is associated with changes in x and changes in y. A particularly important example of a "third factor" is time: if we have x measured for a number of years, and y similarly, then a correlation between x and y could simply be the result of changes that have been happening over the years. For example, ownership of televisions has increased over recent decades, and so has the death rate from lung cancer among women. But no one supposes that owning televisions is causing lung cancer.

- As to testing whether an observed correlation is significantly different from 0, see paragraphs 186–187 below.

- Correlation is a statistic that is sensitive to outliers. It is like the mean and the s.d. in this respect. Just as we may prefer the median and the IQR because they are less affected by outliers, so we may calculate *Spearman's rank correlation* in preference to the product-moment correlation.

- The quantity r^2 is the proportion of variance explained by the regression line, in the following sense:[14]

$$r^2 = \frac{\Sigma\,(\hat{y} - \bar{y})^2}{\Sigma\,(y - \bar{y})^2}. \tag{16}$$

52. **Rank correlation.** The procedure for calculating Spearman's rank correlation is as follows.

- Rank the x's from lowest to highest. (Tied observations receive their average rank.) Rank the y's from lowest to highest. (Again, if there are tied observations, they receive their average rank.) Further calculations are based on these two sets of ranks, not on the original observations.

For the data on linkage of rice markets, the ranks are as in the third and fourth columns below:

---

[14]To obtain this equation, appreciate that

- $\hat{y} - \bar{y} = a + bx - \bar{y} = \bar{y} - b\bar{x} + bx - \bar{y} = b(x - \bar{x})$,
- $S_{xx} = \Sigma\,x^2 - \frac{1}{n}(\Sigma\,x)^2 = \Sigma\,(x - \bar{x})^2$, as in footnote 4 (and similarly for S_{yy}).

Then we see that

$$\frac{\Sigma\,(\hat{y} - \bar{y})^2}{\Sigma\,(y - \bar{y})^2} = \frac{b^2\Sigma\,(x - \bar{x})^2}{\Sigma\,(y - \bar{y})^2} = \frac{S_{xy}^2}{S_{xx}^2}.\frac{S_{xx}}{S_{yy}} = \frac{S_{xy}^2}{S_{xx}S_{yy}} = r^2.$$

x	y	Rank(x)	Rank(y)	d	d^2
5.4	0.2	4	$1\frac{1}{2}$	2.5	6.25
4.0	0.3	2	$3\frac{1}{2}$	−1.5	2.25
8.1	0.2	5	$1\frac{1}{2}$	3.5	12.25
10.0	0.3	6	$3\frac{1}{2}$	2.5	6.25
2.4	0.5	1	5	−4.0	16.00
5.2	0.6	3	6	−3.0	9.00
					52.00

- Calculate the differences d between the x-ranks and the y-ranks. Square these differences. Add up these squared differences.

  As shown above, here this results in a total of 52.

- Calculate Spearman's rank correlation, r_s, from

$$r_s = 1 - \frac{6\,\Sigma\,d^2}{n^3 - n}. \tag{17}$$

  For this data, the result is $1 - \frac{6\times 52}{216-6} = -\frac{102}{210}$, which is -0.49.

- When there are tied observations, as there are in this data, a correction term ought to be added on to $\Sigma\,d^2$. But this is usually quite small, and for simplicity I have omitted it.

(Values of the rank correlation are not directly comparable with values of the product-moment correlation — they are different things, and should not be compared.) Because it is determined by the ranks of the observations, r_s is not so affected by outliers as r is.

Part II. Probability

53. **What is meant by "probability"?** The probability of an event is the chance that it will occur. There are three main ways of refining this idea.

- The frequency interpretation. If we repeatedly give the event the opportunity of occurring, its probability is the proportion of times it actually does occur. Thus, this relies upon empirical data.

    If we say that the probability of a train from Liverpool to Redfern being on time is 0.78, then we must have collected data on the timekeeping of such trains. Perhaps we had records of 50 journeys, found 39 were on time, and calculated $\frac{39}{50} = 0.78$.

- The *a priori* approach.[15] Sometimes, the experimental set-up is so clear, we know the probabilities in advance of collecting any data.

    For an unbiased (fair) coin, the probability of it landing head uppermost equals the probability of it landing tail uppermost, $\Pr\{\text{Head}\} = \Pr\{\text{Tail}\} = 0.5$. When rolling an unbiased six-sided dice, $\Pr\{1\} = \Pr\{2\} = \Pr\{3\} = \Pr\{4\} = \Pr\{5\} = \Pr\{6\} = \frac{1}{6}$. When drawing from a well-shuffled pack of playing cards, $\Pr\{\text{Ace}\} = \frac{4}{52} = \frac{1}{13}$.

[15] The Latin phrase *a priori* in this context means deriving from self-evident propositions.

- Subjective assessments. "I think there is a 30 per cent chance it will rain tomorrow." "There's a 50:50 chance that the Magpies will beat the Roosters." People use language like this. But what does it mean? There is no data, so the frequency interpretation is not relevant. The situation is complex, and the probability obviously cannot be written down on a theoretical basis. A partial answer is that we can imagine collecting data, and being able to make statements like "Of all days at about this time of year, with weather like today's, with weather having been as it has been over the past month, and other relevant factors being similar to those now, for 30 per cent it rained on the following day."

Fortunately (for it is a very difficult subject), we do not need to worry too much about the philosophy of probability. Probabilities are much easier to use in practical calculations than they are to philosophise about.

54. **Rules for doing calculations with probabilities.** Remember this: probabilities are always between 0 (for events that are impossible) and 1 (for events that are certain to occur). So if you get an answer that is supposed to be a probability but is negative or is greater than 1, YOU'VE MADE A MISKATE.[16] Paragraphs 55–61 will be given over to explaining four rules that enable us to do calculations with probabilities.

55. **Addition rule for mutually-exclusive events.** "Mutually-exclusive" means they cannot occur together. Consequently, the probability of one or other of two mutually-exclusive events occurring is the sum of their individual probabilities.

> Suppose Pr{Temperature increases} = 0.62 and Pr{Temperature is unchanged} = 0.23. Then Pr{Temperature increases or is unchanged} $= 0.62 + 0.23 = 0.85$.

The probabilities of all possible exclusive outcomes add up to 1.

[16] If this happens in an examination, and you are in such a rush that you do not have time to correct it, you should make a note that you know you have made a mistake — otherwise, the examiner will think you are a complete idiot.

Suppose Pr{Temperature increases} = 0.62 and Pr{Temperature is unchanged} = 0.23. Then the only other possibility is that the temperature decreases, and Pr{Temperature decreases} = 1 − 0.62 − 0.23 = 0.15.

56. Multiplication rule for independent events. The probability of both happening is the product of their individual probabilities, if the events are independent.

Suppose I roll a fair dice and draw a card from a well-shuffled pack. What happens to the dice does not affect what happens to the cards, and vice versa. The probability of getting a 2 with the dice and a King with the cards is $\frac{1}{6} \times \frac{1}{13} = \frac{1}{78}$.

57. Do not confuse "exclusive" with "independent". Their meanings are very different. (See paragraphs 65–67.)

58. General addition rule. Pr{A or B} = Pr{A}+Pr{B} − Pr{A and B}. That is, the probability of one or the other (or possibly both) equals the sum of the individual probabilities, minus the probability of both occurring. It is standard practice, by the way, for "A or B" to be used to mean "A or B or possibly both".

Suppose we are told that Pr{Rain} = 0.27, that Pr{Wind} = 0.24, and that Pr{Rain and wind} = 0.15. Then we can work out that Pr{Rain or wind} = 0.27 + 0.24 − 0.15 = 0.36. (And we also know that Pr{Neither rain nor wind} = 1 − 0.36 = 0.64.)

59. We can use a Venn diagram to see why the general addition rule holds:

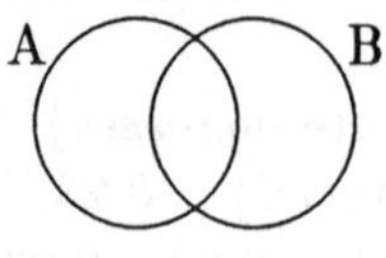

We must subtract off Pr{A and B} because otherwise we would be double-counting the events in the overlapping area.

60. Another way of looking at this is as follows:

The set {A or B} is made up of the three constituents {A but not B}, {A and B}, and {B but not A};

{A} is made up of {A but not B} plus {A and B};

{B} is made up of {B but not A} plus {A and B};

And therefore {A or B} is {A}, plus {B}, minus {A and B}.

(If you know it, the use of set-theory notation is preferable to my notation.)

61. **General multiplication rule.** Pr{A and B} = Pr{A} × Pr{B|A}. (The vertical line | means "given that".) That is, the probability of both A and B occurring is the probability that A occurs multiplied by the probability of B occurring conditional upon A occurring.

Pr{B|A} is referred to as a *conditional probability.*

(And, clearly, Pr{A and B} also is the probability that B occurs multiplied by the probability of A occurring conditional upon B occurring.)

> Suppose that of a group of 12 people, 9 are native-born and 3 are foreign-born. If we select two at random, what is the probability that both are foreign-born? Let A be "the first is foreign-born", and B be "the second is foreign-born". Pr{A} is $\frac{3}{12} = \frac{1}{4}$. Once we know that A has occurred, we know there are 11 remaining of whom 2 are foreign-born. So Pr{B|A} is $\frac{2}{11}$. So Pr{A and B} is $\frac{1}{4} \times \frac{2}{11} = \frac{1}{22}$. (If you know about permutations, combinations, etc., you will recognise that another way to work this out is as $\frac{\text{Number of ways of choosing 2 from 3}}{\text{Number of ways of choosing 2 from 12}} = \frac{3}{66} = \frac{1}{22}$.)

62. **Diagrams that help with probability calculations: Venn diagrams.** Suppose we are told that of a total of 200 students, 77 study accounting, 64 study law, and 92 study neither. We can enter this information on a Venn diagram as follows.

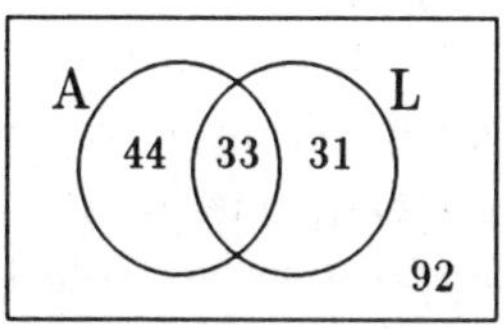

These numbers were worked out as follows.

- 92: we were told this.
- We thus knew the other three numbers had to total $200 - 92 = 108$.
- 44 came from $108 - 64$.
- 31 came from $108 - 77$.
- 33 came from $77 - 44$, or from $64 - 31$.

So if we are asked what is the probability that a randomly-chosen student is studying both accounting and law, we see the answer is $\frac{33}{200} = 0.165$. Or if the question asks what is the probability that a student who is studying accounting does not study law, we see the answer is $\frac{44}{77} = \frac{4}{7}$.

63. We can also use Venn diagrams to represent 3 features, using 3 overlapping circles.

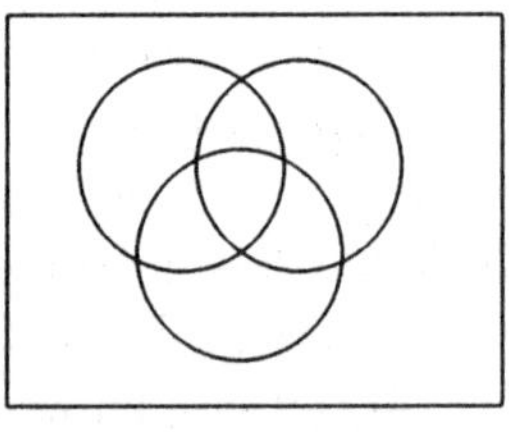

I will not give many further examples of Venn diagram problems because most students get plenty of practice with these in their mathematics course. To answer them, one usually searches the question for a region in the diagram for which the exact number is given, and writes that number on the diagram. Then the numbers for the other regions of the diagram are successively filled in using other pieces of information in the question, by appropriate subtractions. When the diagram is complete, one reads off the answer required.

64. Sometimes, a tricky question is set, in which a vital piece of information is missing. Instead, one is informed that two of the features are independent. Suppose we are told that of a total of 200 students, 80 study biology, 90 study neither biology nor geography, and the choice of whether a student does or does not study biology

is independent of their choice of studying or not studying geography. Then the Venn diagram is filled in as follows.

- 90 are outside both B and G.
- Since we know 80 are inside B, the total outside B is $200 - 80 = 120$.
- The number outside B but inside G must therefore be $120 - 90 = 30$.
- Seeing the 120 outside B are split 30:90 according to whether they are inside G or outside, we use the independence property to deduce that the 80 inside B must be split 20:60 (that is, in the same proportions) as to whether they are inside G or outside.

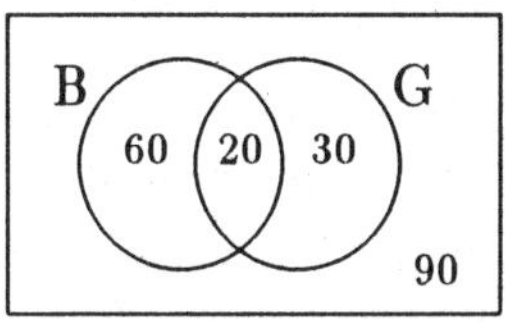

65. **Note on "exclusive" and "independent".** It is very important that the student understands the distinction between these two words.

- "Exclusive" events are ones that can never occur together, i.e., A never occurs when B occurs.
- "Independent" events are ones where the occurrence or nonoccurrence of one is unconnected with the occurrence or nonoccurrence of the other, i.e., the proportion of times A occurs is the same whether or not B occurs.

I think some students have difficulty because in the English language the word "independence" carries with it a connotation of separateness. This can be misleading here: when A and B are independent, it is their *causes* that are separate; but all possible combinations of the various possibilities (A and B, A but not B, B but not A, neither A nor B) may be observed.

66. **"Exclusive" and "independent" on a Venn diagram.** If the numbers in the four regions of a Venn diagram are w, x, y, z, as shown below,

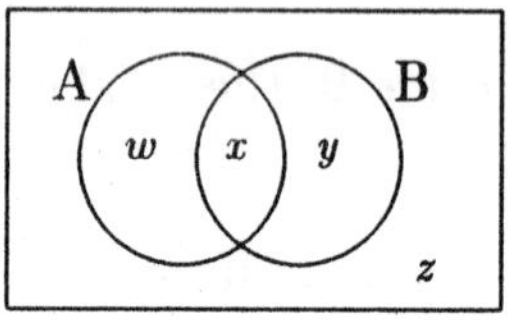

then

- A and B are mutually exclusive if $x = 0$;
- A and B are independent if $\frac{w}{x} = \frac{z}{y}$ (notice that w and x refer to events that are inside A, and y and z refer to events that are outside A: $\frac{w}{x}$ is $\frac{\text{outside B}}{\text{inside B}}$, and $\frac{z}{y}$ is also $\frac{\text{outside B}}{\text{inside B}}$).

67. To further clarify the distinction between "exclusive" and "independent", consider the following example. A certain mineral ore contains two radioactive elements. One emits α-particles, the other emits β-particles. The radioactive decay of any atom is completely unaffected by that of other atoms. Suppose we have an instrument that detects all particles emitted, α and β.

- Whether at least one α-particle is detected or not in any given time period is independent of whether at least one β-particle is detected or not.
- However, suppose we focus our attention on the particles, and classify them as either α or β. Now the two alternatives are mutually exclusive. In effect, what we are doing by focussing on the particles is restricting ourselves to the two areas of the Venn diagram, $\{\alpha$ but not $\beta\}$ and $\{\beta$ but not $\alpha\}$.

68. **Diagrams that help with probability calculations: complete listing of all possibilities.** Suppose we are asked what is the probability of getting a total of at least 8 and at most 10 when rolling two dice. We write out a table of all the possible combinations of outcomes, with the totals shown:

		Second dice 1	2	3	4	5	6
	1	2	3	4	5	6	7
First	2	3	4	5	6	7	*8*
	3	4	5	6	7	*8*	*9*
	4	5	6	7	*8*	*9*	*10*
dice	5	6	7	*8*	*9*	*10*	11
	6	7	*8*	*9*	*10*	11	12

I have indicated which outcomes satisfy the required condition ("a total of at least 8 and at most 10") by printing them in italics. To find the answer to the question, we simply add up the probabilities of the outcomes which satisfy the required condition. For this example, each of these probabilities is $\frac{1}{6} \times \frac{1}{6} = \frac{1}{36}$. There are 12 of them, so the answer is $\frac{12}{36} = \frac{1}{3}$.

68x. **The probability of Venus.** The randomisation device used in ancient times, before the invention of dice, was the *astragalus*. This is a small bone from an animal's ankle, and, as compared with the six-sided dice that we are familiar with, is decidedly irregular: because of the shape, there are only four sides that can come uppermost, and these do so with probabilities of approximately 0.1, 0.4, 0.4, and 0.1; the respective scores are 1, 3, 4, and 6 (David, 1955). The best of all throws with four astragali was when the four sides were all different (this was termed "Venus"). To work out the probability of this, we start off by trying to list all the possibilities:

A1	A2	A3	A4	Probability
1	1	1	1	$0.1 \times 0.1 \times 0.1 \times 0.1$
1	1	1	3	$0.1 \times 0.1 \times 0.1 \times 0.4$
1	1	1	4	$0.1 \times 0.1 \times 0.1 \times 0.4$
1	1	1	6	$0.1 \times 0.1 \times 0.1 \times 0.1$
1	1	3	1	$0.1 \times 0.1 \times 0.4 \times 0.1$
1	1	3	3	$0.1 \times 0.1 \times 0.4 \times 0.4$
⋮	⋮	⋮	⋮	⋮

and so on. At about this point, we realise (a) that there are rather a lot (actually, $4^4 = 256$) of possibilities, and (b) we only need to write out the combinations constituting Venus:

A1	A2	A3	A4	Probability
1	3	4	6	$0.1 \times 0.4 \times 0.4 \times 0.1 = 0.0016$
1	3	6	4	$0.1 \times 0.4 \times 0.1 \times 0.4 = 0.0016$
1	4	3	6	$0.1 \times 0.4 \times 0.4 \times 0.1 = 0.0016$
⋮	⋮	⋮	⋮	⋮

and so on. We further realise (c) each of these combinations occurs with the same probability, namely, 0.0016, and (d) the number of such combinations is 24.

> A1 can be any one of four possibilities, A2 must be one of the three possibilities different from A1, A3 must be one of the two possibilities different from both A1 and A2, and A4 must be the possibility not occurring yet, making a total of $4 \times 3 \times 2 \times 1 = 24$ combinations.

The probability of throwing Venus is therefore $24 \times 0.0016 = 0.0384$.

69. **Diagrams that help with probability calculations: tree diagrams.** Suppose we examine a ceramic bowl that appears to be of 15th-century origin. Our experience has been that 25 per cent of bowls like this one are in fact 19th-century fakes. We have a method of testing, but it is a fallible one — that is, it sometimes makes mistakes. Specifically, 92 per cent of fake bowls fail the test, but 16 per cent of genuine ones also fail. What is the probability that a bowl that is found to fail the test is truly a fake? To answer this, we present the information in the form of a tree diagram.

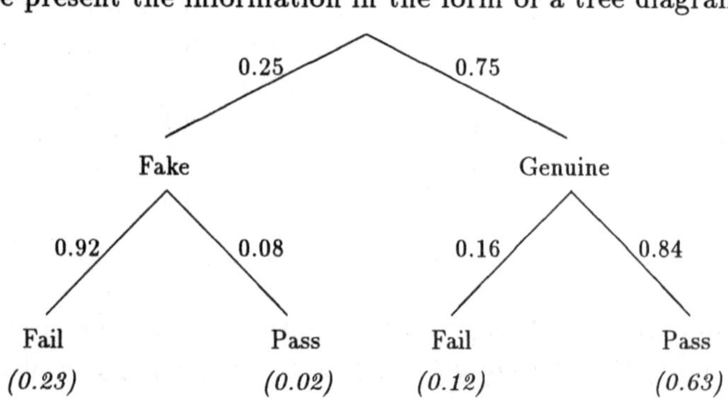

(Of course, the 0.75 came from $1-0.25$, the 0.08 came from $1-0.92$, and the 0.84 came from $1-0.16$.) At the bottom of the tree diagram, the probabilities of the different outcomes are shown: $0.25 \times 0.92 = 0.23$, $0.25 \times 0.08 = 0.02$, $0.75 \times 0.16 = 0.12$, and $0.75 \times 0.84 = 0.63$. And now we are ready to determine the answer. The probability of a randomly-selected bowl failing the test is $0.23 + 0.12 = 0.35$; of this total, a contribution of 0.23 came from bowls that actually were fake; the required answer is therefore $\frac{0.23}{0.35} = 0.66$. Notice that the method of calculation is essentially the use of the general multiplication rule: we are calculating Pr{Fake|Fail test} as $\frac{\Pr\{\text{Fake and fail test}\}}{\Pr\{\text{Fail test}\}}$.

Students sometimes find this sort of question easier if they think in terms of the *numbers* of events we would expect to observe, instead of the *probabilities*. For instance, suppose we started with 1000 bowls. Then we would expect to find that 350 of them fail the test; of these, 230 really should have failed. So the required answer is $\frac{230}{350} = 0.66$.

70. This is a common type of problem — sometimes phrased in medical language (having or not having a disease, being diagnosed as having it or not), sometimes phrased in legal language (truly guilty or innocent, being convicted or acquitted), sometimes phrased in other language (e.g., suitable for employment or not, passing a selection test or failing). In answering it, it is crucial to keep clear the distinction between what the test is saying and what the state of affairs truly is. The point about such problems is that in practice one knows the test result, and wants to deduce what is the most likely true state of affairs.

Notice that one needs a very accurate test if the condition to be detected is a rare one. Suppose that a disease has an incidence of 0.1 per cent; that the probability of getting a positive test result from a diseased person is 0.99; and that the probability of getting a positive result from a healthy person is 0.02. Then the probability that a person who gives a positive test result really does have the disease is $\frac{0.001 \times 0.99}{(0.001 \times 0.99) + (0.999 \times 0.02)}$, which works out to be only 0.047. Hence the need for further tests and investigations following a first positive test.

71. **Asking sensitive questions in surveys.** An interesting class exercise to reinforce your grasp of probability calculations and to introduce you to a subject of considerable practical importance is the following. Suppose I want to carry out a survey to find what proportion of the population drive faster than the speed limit, fail to pay taxes, go to church on Sunday, are in favour of stricter gun laws, etc. I am not the least bit interested in whether *you personally* drive faster than the speed limit, but I do want to get an honest overall estimate for the whole population — and people are often not honest when answering questions like these. The *randomised response* method can be used here. It relies upon the respondent knowing that the interviewer does not know whether the respondent is answering the sensitive question, or is answering a perfectly harmless question. An example:

- Spin a coin twice. Show no-one the results.
- If the first spin resulted in a head, answer the question marked H. If the first spin resulted in a tail, answer the question marked T.

  H: Outside of class, do you study for at least 15 hours per week?

  T: Did the second spin of the coin result in a tail?

- Let p be the proportion of students who study for at least 15 hours per week outside of class. The tree diagram is as follows:

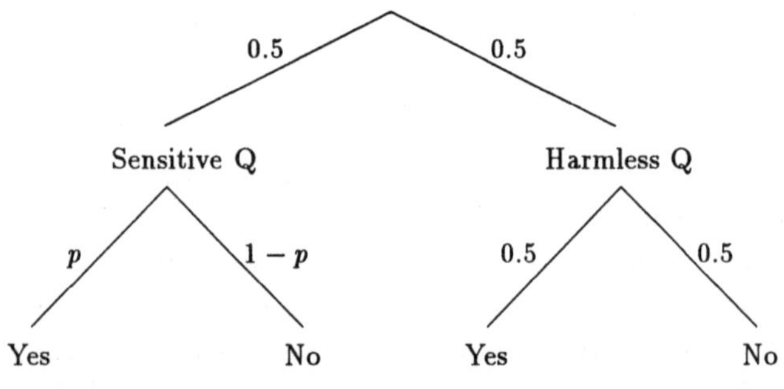

- Let the proportion of people who answer "yes" be y. This is what I can find in my survey. From the above diagram, I can see that it is also $(0.5 \times p) + (0.5 \times 0.5)$, which is $0.5p + 0.25$. Knowing that $y = 0.5p + 0.25$, I can find p from the equation $p = 2y - 0.5$. For example, in a small class of 35 students, 12 answered "yes". I deduce that $p = 2 \times \frac{12}{35} - 0.5$, which is 0.19.

72. **An example of a probability calculation.** Suppose that the diagram below represents part of an electric circuit.

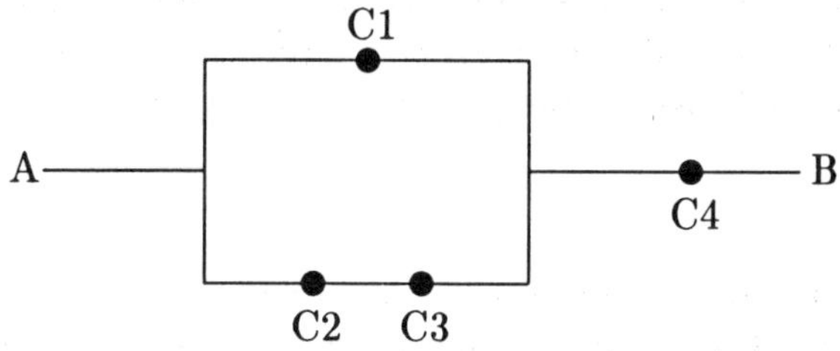

Current only flows from A to B if

- Component C4 functions, and
- Either
  - Component C1 functions, or
  - Both C2 and C3 function.

The four components behave independently;[17] the probability of any one of them functioning is p. Determine the probability that current flows.

73. We know

Pr{C1 functions} $= p$, and

[17] A comment about the "independence" assumption is worth making here. Many systems for which safety is important are designed so that failure of one component does not lead to system failure — the simultaneous failure of several components is required for this to occur. Each component individually is highly reliable, hence the independent simultaneous failure of several of them is almost unheard-of. Yet failure of chemical manufacturing plants, nuclear power stations, jet aircraft, etc., does occasionally occur. Frequently, it is found that this comes about because the simultaneous failures of the several components occurred *non-independently.* Imagining how this could happen and preventing it is an important part of safety engineering.

Pr{Both C2 and C3 function} = p^2 (because we are told C2 and C3 behave independently).

Consequently,

Pr{C1 functions, or both C2 and C3 function} = Pr{C1 functions} + Pr{Both C2 and C3 function} − Pr{C1 functions and both C2 and C3 function} (this is using the general addition rule), $= p + p^2 - p \times p^2$ (we use independence again to get the $p \times p^2$ term), $= p(1 + p - p^2)$.

Finally,

Pr{C4 functions and the left hand part of the circuit functions} $= p \times p(1 + p - p^2)$ (again using the independence property), $= p^2(1 + p - p^2)$,

which is the required answer.

74. If you get too hopelessly confused with this type of question, it may be practicable to list all possible combinations of what can happen (if there are not too many), determine for each what the outcome is, and add up the probabilities of the combinations for which the required outcome occurs. In the table which follows, 0 is used to indicate the component fails, and 1 to indicate it functions.

C1	C2	C3	C4	Current?	Probability
0	0	0	0	No	
0	0	0	1	No	
0	0	1	0	No	
0	0	1	1	No	
0	1	0	0	No	
0	1	0	1	No	
0	1	1	0	No	
0	1	1	1	Yes	$(1-p)p^3$
1	0	0	0	No	
1	0	0	1	Yes	$(1-p)^2p^2$
1	0	1	0	No	
1	0	1	1	Yes	$(1-p)p^3$
1	1	0	0	No	
1	1	0	1	Yes	$(1-p)p^3$
1	1	1	0	No	
1	1	1	1	Yes	p^4

Adding up the probabilities in the final column, we get $(1-p)^2p^2+3(1-p)p^3+p^4 = p^2[(1-p)^2+3p(1-p)+p^2] = p^2(1+p-p^2)$, as before.

75. **Probability distributions.** A probability distribution describes the probabilities with which all the different possible outcomes occur.

- It may be something very simple:

Number of heads in one spin of a coin:	0	1
Probability:	$\frac{1}{2}$	$\frac{1}{2}$

- It may be something a little more complicated, with some particular pattern to it:

Number of heads in four spins of a coin:	0	1	2	3	4
Probability:	$\frac{1}{16}$	$\frac{4}{16}$	$\frac{6}{16}$	$\frac{4}{16}$	$\frac{1}{16}$

(For how these probabilities were calculated, see paragraph 84 below.)

- There may be no theoretical pattern to the probabilities:

Number of people in a car:	1	2	3	4+
Probability:	0.43	0.36	0.15	0.06

- Remember that probabilities are always between 0 and 1, and that they must add up to 1 if we have a list of mutually exclusive and exhaustive events (as we do in this context).

76. In the above examples, the outcomes were *discrete*, and they consisted of *counts* — we could have 0 or 1 or 2 or ... heads, or people in a car, but we could not have 1.5 or 0.73. Contrast this with

continuous variables — measurements like 5.328 kilograms, or 26.7 degrees Celsius. These may be rounded to the nearest kilogram or degree for the purpose of presenting and communicating the information, but in principle there is nothing special about whole numbers here. In this case, we have a *probability density function*. For this, see paragraph 94.

77. **The binomial distribution.** This is one of the most important discrete probability distributions. To introduce it, suppose we have a six-sided dice, with four sides painted blue and two sides painted red. The dice is an unbiased one, so at any roll, there is a one-third chance of getting a red face uppermost, and a two-thirds chance of getting a blue face uppermost.

Tree diagram for one roll:

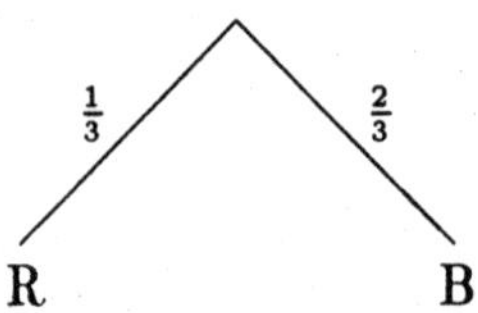

The corresponding probability distribution of the number of blue faces coming uppermost is:

Number of blues:	0	1
Probability:	$\frac{1}{3}$	$\frac{2}{3}$

Tree diagram for two rolls:

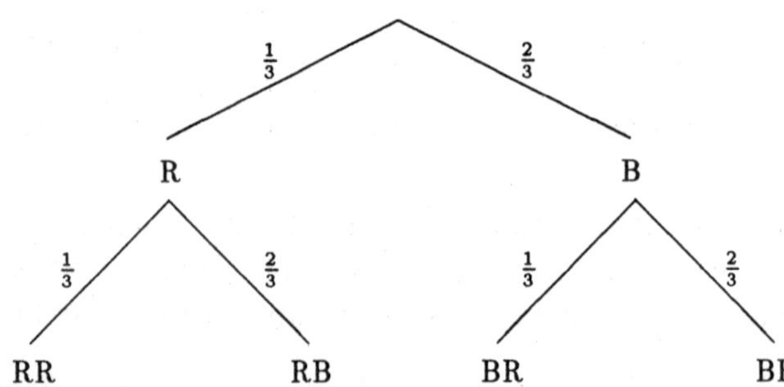

Now, the probability distribution of the number of blue faces coming uppermost is:

Number of blues:	0	1	2
Probability:	$\frac{1}{9}$	$\frac{2}{9}+\frac{2}{9}$	$\frac{4}{9}$
		$=\frac{4}{9}$	

Tree diagram for three rolls:

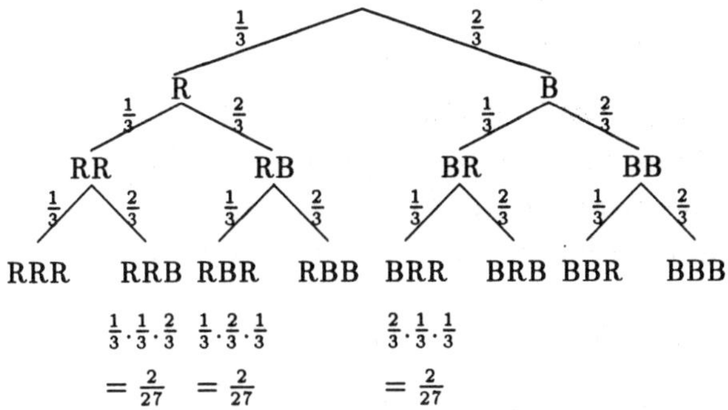

To illustrate how the probabilities are calculated, the case of getting one blue face uppermost has been shown. It occurs in three ways; for each, the probability is $\frac{2}{27}$. The complete probability distribution of the number of blue faces coming uppermost is:

No. of blues:	0	1	2	3
Probability:	$\frac{1}{27}$	$\frac{2}{27}+\frac{2}{27}+\frac{2}{27}$	$\frac{4}{27}+\frac{4}{27}+\frac{4}{27}$	$\frac{8}{27}$
		$=\frac{6}{27}$	$=\frac{12}{27}$	

78. Notice these four things:

- Each of the probability distributions refers to one particular level of the tree diagram: the first to level 1 (one roll), the second to level 2 (two rolls), and the third to level 3 (three rolls).

- There are only two alternatives at each branching point, and they are always the same two.
- The same pair of probabilities for the two alternatives always occurs, at whatever level we are in the tree diagram.
- The same pair of probabilities for the two alternatives always occurs, whatever has been the outcome at previous levels of the tree diagram.

Of course, to get the probability at an end of the tree, we multiply the probabilities applying to the branches leading to it.

79. These are, indeed, the crucial features of the conditions for the binomial distribution to occur. Rewording them,

- There is a particular number (n) of *trials*.
- At each trial, one or other of *two* possibilities occurs, and they are always the same two. (The two possibilities are traditionally referred to as "success" and "failure".[18])
- The probability of a success stays *constant* from trial to trial. The symbol p is used for it, so $1 - p$ is the probability of a failure.
- The results of the different trials are *independent* of each other.

80. The binomial distribution is concerned with the number of successes that occur in the n trials. It enables us to calculate the probability of obtaining exactly x successes, when the probability of success at any one trial is p. The expression for it is:

$$\binom{n}{x} p^x (1-p)^{n-x} \tag{18}$$

(provided x is a whole number between 0 and n, inclusive). The symbol $\binom{n}{x}$ ("n choose x") means the number of ways of choosing

[18] But these names are arbitrary, and it may be convenient in some particular application to refer to something bad, e.g., someone being struck by disease, or a river flooding, as being a "success".

x objects from a set of n objects.[19] The logic behind this formula is that the probability that a *particular set* of x trials all result in successes, with the remaining $n - x$ trials resulting in failures, is $p^x(1-p)^{n-x}$ (because the outcomes of the trials are independent, and so we can multiply probabilities); and there are $\binom{n}{x}$ different sets of x trials that can be found among the total n trials.

81. Buddhist priests in 19th-century Tibet made reincarnation predictions by throwing a six-sided dice with letters on the faces onto a rebirth chart. It has been alleged that the dice may have been loaded so as to frequently show the letter Y, which predicts a ghostly existence in the future life, and thus necessitates the performance of many expensive rites to counteract so undesirable a fate (David, 1955). Suppose this is so, and for the purpose of this example, let the probability of throwing a Y be 0.3, rather than the $\frac{1}{6}$ that it would be if the dice was honest. Suppose also that the priest throws the dice for six people. Find the probabilities of Y occurring 0, 1, 2, 3, 4, 5, and 6 times in the 6 throws, assuming that the conditions for the binomial distribution hold.[20]

> We need a convenient notation to represent the probability of a given number of Y's. A commonly-used notation is $\Pr\{X = x\}$. Be sure to understand the meaning of this — the capital X

[19] The formula for calculating it is

$$\binom{n}{x} = \frac{n!}{x!(n-x)!},$$

where $x!$ is "x factorial", $x! = x(x-1)(x-2)\ldots(3)(2)(1)$. Similarly, $n! = n(n-1)(n-2)\ldots(3)(2)(1)$ and $(n-x)! = (n-x)(n-x-1)\,(n-x-2)\ldots(3)(2)(1)$. (A special case is that 0! is defined to be 1.) For example,

$$\begin{aligned}\binom{8}{3} &= \frac{8.7.6.5.4.3.2.1}{3.2.1.5.4.3.2.1}\\ &= \frac{8.7.6}{3.2.1} \quad \text{(notice it has been possible}\\ &\quad\ \text{to cancel out a large number of terms)}\\ &= 56.\end{aligned}$$

Notice also that $\binom{n}{0} = \binom{n}{n} = 1$, whatever value n is. And $\binom{n}{1} = \binom{n}{n-1} = n$, whatever value n is.

[20] For example, that whether the dice does or does not show Y at one throw is independent of what it shows at the previous throws.

stands for the name of the variable we are concerned with (e.g., the number of Y's), and little x stands for a particular number (e.g., 2); it may also be helpful to read the symbol = as "takes the value", rather than as "equals".

$$\begin{aligned}
\Pr\{X=0\} &= \tbinom{6}{0}0.3^0 0.7^6 \\
&= 1 \times 1 \times 0.118 \quad \text{(remember that any number raised to the power 0 results in 1)} \\
&= 0.118 \\
\Pr\{X=1\} &= \tbinom{6}{1}0.3^1 0.7^5 \\
&= 6 \times 0.3 \times 0.168 \\
&= 0.303 \\
\Pr\{X=2\} &= \tbinom{6}{2}0.3^2 0.7^4 \\
&= 15 \times 0.09 \times 0.240 \\
&= 0.324 \\
\Pr\{X=3\} &= 0.185 \\
\Pr\{X=4\} &= 0.060 \\
\Pr\{X=5\} &= 0.010 \\
\Pr\{X=6\} &= 0.001
\end{aligned}$$

(The easiest probabilities to work out are $\Pr\{X = 0\}$ and $\Pr\{X = n\}$. The first of these, the probability of no successes, is Pr{Failure at trial 1} × Pr{Failure at trial 2} × ... × Pr{Failure at trial n} $= (1-p)^n$. The second of these, the probability of all successes, is Pr{Success at trial 1} × Pr{Success at trial 2} × ... × Pr{Success at trial n} $= p^n$. These results are evident even without knowing the binomial formula.)

82. It is now straightforward to answer questions like "what is the probability that at most 2 Y's occur?" (The answer is $0.118 + 0.303 + 0.324 = 0.745$.) The chief reason for getting such questions wrong is not paying sufficiently careful attention to the exact wording — look for whether it refers to "at least", "at most", "more than", "less than", "not less than", "not more than", or whatever.

83. The probabilities can be presented in the form of a histogram:

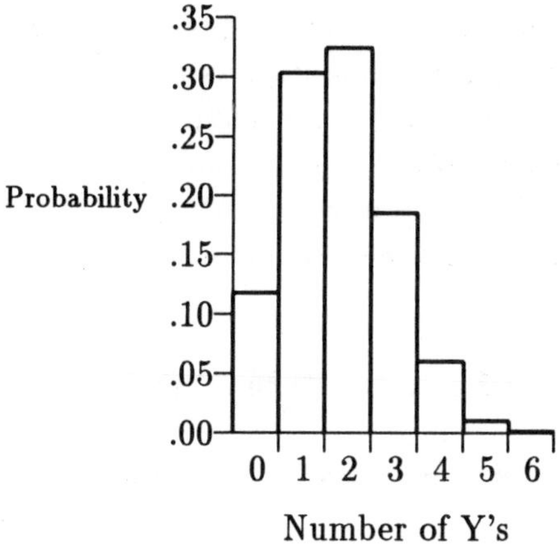

84. The second example in paragraph 75 is the binomial distribution with $n = 4$ and $p = \frac{1}{2}$.

85. **Properties of the binomial distribution.** The mean and s.d. of the number of successes are:

$$\text{mean} = np \tag{19}$$

$$\text{s.d.} = \sqrt{np(1-p)}. \tag{20}$$

(The first of these formulae is just a matter of common sense; regarding the second, see paragraph 124.) Further, the variance — always the square of the s.d. — is $np(1-p)$.

86. **The Poisson distribution.** Another important discrete distribution is the Poisson distribution.[21] There is a close relationship with the binomial distribution, as follows. In the binomial situation, imagine that the number of trials n gets larger and larger, the probability of success p gets smaller and smaller, and that this happens in such a way that the mean number of successes np stays constant. Then the limiting distribution (as $n \to \infty$) is the Poisson distribution.

[21]It is named after the French mathematician Siméon-Denis Poisson (1781–1840).

Notice that having infinite n means we have effectively got a continuum, not a finite number of isolated trials; and whilst we can count up the number of success events that occur in any stretch of the continuum, it is meaningless to try to count the failure events — there are an infinite number of them. With the Poisson distribution, the usual terminology is to refer to the number of *events* that happen, rather than the number of *successes*. The following diagram should make the contrast plain.

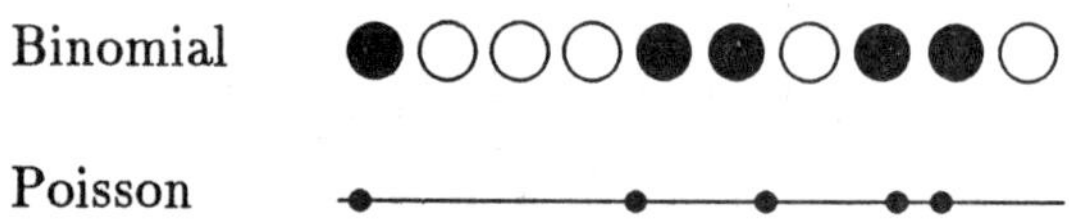

With the binomial distribution, either of two types of event (i.e., success or failure) are occurring at particular opportunities (i.e., trials). With the Poisson distribution, the relevant event is occurring at random points on a continuum. (In many practical applications, the continuum is of time, and in many other applications, it is one of length.)

87. In the case of the Poisson distribution, the expression for the probability of exactly x events is

$$\frac{\lambda^x e^{-\lambda}}{x!} \tag{21}$$

(provided x is a whole number greater than or equal to 0), where λ is the average number of events (analogous to np in the binomial distribution).

In many books, λ is split up as $\lambda = \gamma t$, where γ is the rate of events per unit time and t is the time period being considered (or, in other contexts, γ is the density of events per unit distance and t is the length being considered). The symbols λ and γ are the Greek letters lambda and gamma.

The mathematical derivation of (21) from (18) is beyond the scope of this book.

88. The Poisson distribution, or the Poisson *process* that gives rise to it, is often viewed as embodying complete randomness: at each point on the continuum, there is the same probability of an event occurring there,[22] and whether an event does occur there or not is entirely independent of where events are occurring elsewhere on the continuum.

89. For an example, suppose wars begin at a rate of 0.7 per year. What is the probability that within a 5-year period, exactly 3 wars start? Here, the average number of events over the period being considered is 3.5. (This will be obvious to many readers. Others will have to notice that γ is 0.7 per year, t is 5 years, and work it out as $\lambda = \gamma t = 0.7 \times 5 = 3.5$; naturally, when doing a calculation like this, γ and t have to be in compatible units.) The question asks about $x = 3$. The probability is $\dfrac{3.5^3 e^{-3.5}}{3!}$, which works out to be 0.216.

> If we are asked about the probability of there being 3 or fewer wars, we work out the probabilities of 3, 2, 1, and 0 wars, and add them together.
>
> If we are asked about the probability of more than 3 wars, we work out the probabilities of 3, 2, 1, and 0 wars, add them together, and then subtract the total from 1.

For statistical studies of the occurrence of wars, see Richardson (1944) and Cioffi-Revilla (1991). To say that, when looked at in the large, the occurrence of war resembles a Poisson process, is perfectly compatible with the conventional view that specific wars have specific causes, the study of which is of proper concern to the historian. Richardson (1944) gives this homely analogy: "There are similar contrasts in other social affairs; the statistics of marriage, for example, are in contrast with any biographer's account of the incidents that led two named persons to marry each other."

---

[22]You may have noticed that this is loosely worded: strictly speaking, events are occurring at *points*, and there is zero chance of an event occurring *exactly* at any specified location. So it would be more precise to say that at each point on the continuum, there is the same probability of an event occurring within a distance δ of the point, and that this is true no matter how small δ is.

90. **Properties of the Poisson distribution.** The mean and standard deviation of the number of events are:

$$\text{Mean} = \lambda \tag{22}$$
$$\text{s.d.} = \sqrt{\lambda} \tag{23}$$

(and so the variance is λ). Notice that for this distribution, the s.d. is determined by what the mean is: specifically, s.d. = $\sqrt{\text{mean}}$.

> So if we know the annual number of road deaths in a country is about 2000, we expect the s.d. to be about $\sqrt{2000} = 45$, and we are not surprised if the number of deaths this year is one or two s.d.'s different from the number last year. See also paragraph 121 below, on the s.d. of the difference between two random variables. (The vast majority of road deaths occur one at a time, not 10 or 20 at a time as in a very few bus crashes, and the Poisson distribution is a natural choice to make here. But, even in the absence of identifiable causes such as changes in legislation, the annual number actually varies by more than the Poisson distribution predicts; the reasons include the variation in the weather from year to year, and the variation in the level of law enforcement.)

91. **Comparison of the binomial and Poisson distributions.** The following example allows the binomial and Poisson distributions to be compared, both as to their realism and as to the numerical results. Traffic on a road is passing a point at a rate of 300 vehicles per hour. Determine the probability that exactly 4 vehicles pass in a period of 30 seconds.[23]

- To use the binomial distribution, we will say that vehicles do not travel nose-to-tail, and anyway they have some physical length. So after a vehicle has passed, there is a short period of

[23] This sort of calculation will be important if, for example, we are choosing the settings of traffic signals (that is, how long they will stay green, and how long they will be red), and there is only limited road space for the vehicles to wait. We might actually want to know the probability of getting 5 *or more* vehicles, but obtaining the probability of *exactly* x vehicles (for $x = 0, 1, 2, 3, 4$) is clearly a step to the solution.

time during which another vehicle cannot come. Moreover, the road has a finite capacity for traffic flow; for a single lane, this is typically about 1800 vehicles per hour. So, imagine that time is made up of 2-second chunks, in each of which either a vehicle passes (probability p) or does not (probability $1 - p$). The value of p will be 0 for an empty road, 1 for a road operating at capacity, and is $\frac{1}{6}$ in our particular example. There are 15 of these chunks in the 30-second period. Consequently, we work out $\binom{15}{4}(\frac{1}{6})^4(\frac{5}{6})^{11} = \frac{15.14.13.12}{4.3.2.1} \times 0.000772 \times 0.135 = 0.142$.

- To use the Poisson distribution, we will say that vehicles even in a single lane of road can and do travel much closer together than 2 seconds; and, anyway, we might be concerned with a multi-lane road, where vehicles can be as close together in time as they like. Here, λ is 2.5 (300 vehicles per hour is 5 per minute, or 2.5 per 30 seconds). We work out $\dfrac{2.5^4 e^{-2.5}}{4!} = 0.134$.

The two results are close together, but not identical.

92. Both the binomial and the Poisson distributions use the ideas of a constant probability of an event occurring, and of independence between all the opportunities for the event to occur. They differ in that the binomial distribution refers to a finite number of distinct trials, whereas the Poisson distribution refers to occurrences at arbitrary points in a continuum (see paragraph 86 above).

93. **The Poisson distribution as an approximation to the binomial.** In paragraph 86, the Poisson distribution was introduced by saying it came from the binomial when we let

Number of trials $n \to \infty$,

Probability of success $p \to 0$,

in such a way that

The expected number np stays constant $(= \lambda)$.

So if n is "large" and p is "small", the binomial and Poisson distributions will be much the same, and the Poisson may be regarded as an approximation to the binomial. The motivation for doing this is that $\binom{n}{x}$ may be difficult to calculate when n is large and x is not close to either 0 or n. Books often say that the approximation is pretty good if n is at least 100, and np is less than 10.

> For example, if $n = 300$ and $p = 0.007$, the probability of exactly 6 successes is $\binom{300}{6}0.007^6 0.993^{294} = 0.0144$ using the binomial distribution, or is $\dfrac{2.1^6 e^{-2.1}}{6!} = 0.0146$ using the Poisson approximation.

For the use of the *normal* distribution to approximate the binomial, see paragraphs 108–109 below.

94. **Introduction to continuous distributions.** We have already contrasted, in paragraph 76, continuous data with discrete data. We mentioned there the *probability density function* (p.d.f.); its crucial property is that the area beneath it between any two values (x_1 and x_2, say) gives the probability of getting an observation between those two values.

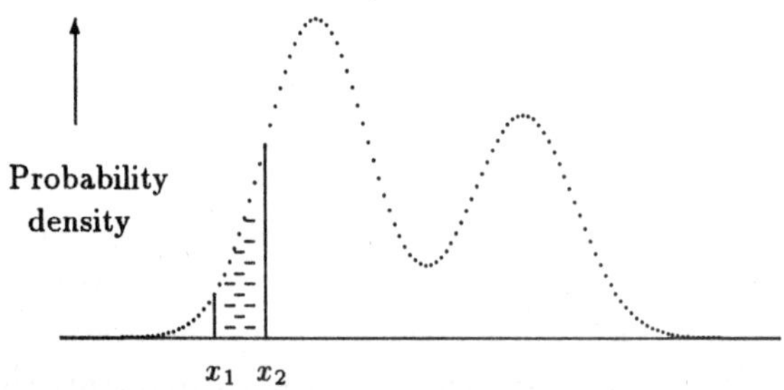

The probability of getting a value in the range x_1 to x_2 is the area marked with dashes.

Notice that:

- The curve cannot be negative (because probabilities cannot be negative).
- The total area under the curve must be 1 (this is just saying that the total probability of a set of mutually exclusive and exhaustive events is 1: some value of x between $-\infty$ and $+\infty$ must be observed).
- There is zero probability that we get any value x *exactly*; so when referring to continuous variables, we need not worry

about distinguishing between being "less than" x and being "less than or equal to" x.

The p.d.f. can be thought of as arising when making histograms of larger and larger sets of data with finer and finer classes of the x-values.

95. The integral of the p.d.f. from $-\infty$ to x is known as the *cumulative distribution function.* It gives the probability of getting an observation less than or equal to x. The shape of it has these properties:

- It is always increasing (if the slope was ever negative, that would mean the p.d.f. was negative, which is impossible).
- It is 0 at $-\infty$.
- It is 1 at $+\infty$.

96. Naturally, we get the probability of obtaining a value between x_1 and x_2 (where x_2 is greater than x_1) from the cumulative distribution at x_2 minus the cumulative distribution at x_1: that is, by integrating the p.d.f. between x_1 and x_2. If $x_2 - x_1$ is very small, this integral will be approximately $(x_2 - x_1) \times$ the p.d.f. at $[x_1 + \frac{1}{2}(x_2 - x_1)]$.

97. **The uniform distribution over the range 0 to 1.** The "uniform" distribution has this name because the probability density is constant throughout the range of possible values of the variable (and is zero elsewhere).

- Illustration of the probability density:

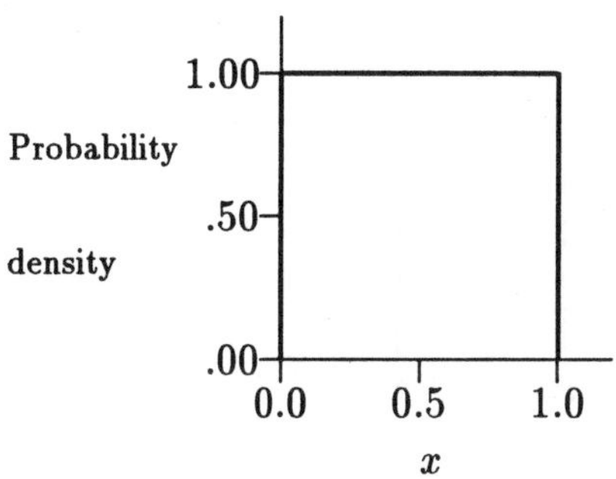

- Expression for the probability density:

$$\begin{cases} 1 & \text{(provided } x \text{ is between 0 and 1)} \\ 0 & \text{(for other } x\text{)} \end{cases}$$

- Thus with this distribution, only values of x between 0 and 1 are possible.

- Expression for the cumulative probability distribution:

$$\begin{cases} 0 & \text{(for } x < 0\text{)} \\ x & \text{(provided } x \text{ is between 0 and 1)} \\ 1 & \text{(for } x > 1\text{)} \end{cases}$$

(Another way of writing this is as $\max(0, \min(1, x))$, and perhaps the reader can think of another.)

- So the probability of getting a value between x_1 and x_2 is simply $x_2 - x_1$ (provided that $0 \le x_1 \le x_2 \le 1$).

- Most statistical software packages have the facility to generate random numbers from this distribution; many statistics books have tables of random whole numbers in them, and these tables can be adapted to the same purpose. The third and fourth examples in Table 4 will show how random numbers from this distribution can be used as the starting point when requiring random observations from some other continuous distribution.

98. The uniform distribution over the range a to b.

- Illustration of the probability density:

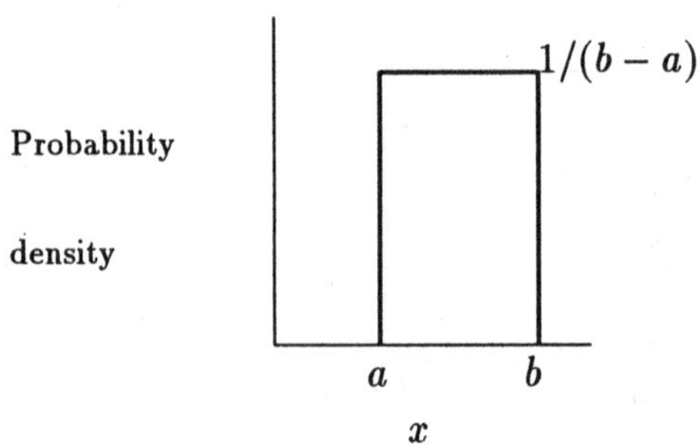

The height of the p.d.f. has to be $1/(b-a)$ for the following reason: the total area under a p.d.f. is always 1; the base of the rectangle is $b-a$; knowing that base $\times$ height $= 1$, we find height $= 1/(b-a)$.

- Expression for the probability density:

$$\begin{cases} 1/(b-a) & \text{(provided } x \text{ is between } a \text{ and } b) \\ 0 & \text{(for other } x) \end{cases}$$

- And the probability of getting a value between x_1 and x_2 is $(x_2 - x_1)/(b-a)$ (provided that $a \le x_1 \le x_2 \le b$).

99. The exponential distribution.

- The expression for the probability density is:

$$\left\{ \begin{array}{ll} \frac{1}{\theta}e^{-x/\theta} & \text{(provided } x \text{ is positive)} \\ 0 & \text{(for negative } x) \end{array} \right\} \qquad (24)$$

- And so the shape looks like this:

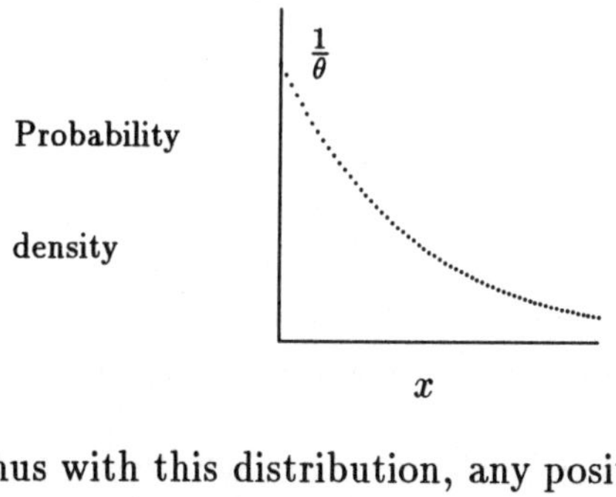

- Thus with this distribution, any positive value of x is possible, but negative values are impossible.

- The expression for the cumulative probability distribution is:

$$\left\{ \begin{array}{ll} 1 - e^{-x/\theta} & \text{(provided } x \text{ is positive)} \\ 0 & \text{(for negative } x) \end{array} \right\} \qquad (25)$$

- The average value of X is θ; this will be demonstrated as the final example in paragraph 115. It is also shown there that the s.d. is θ.

- One reason for the importance of this distribution is that if events are happening in time according to a Poisson process, the times between successive events have an exponential distribution. And if the rate of the Poisson process is γ, the θ of the exponential distribution is $1/\gamma$.

  To show this, appreciate that the probability of 1 or more events occurring within a time t is $1 - e^{-\gamma t}$. (Replace λ by γt in expression (21); set $x = 0$; subtract from 1.) And the probability of 1 or more events occurring between time 0 and time t is the probability that the time elapsing from time 0 until the first event is less than t. Comparing $1 - e^{-\gamma t}$ with expression (25), we see $\gamma = 1/\theta$.

  Notice that the Poisson distribution refers to the count of the number of events in a time period t, whereas the exponential distribution refers to the length of time between one event and the next.

- The distribution is important wherever it is thought that a Poisson process might be occurring. In particular, it is important in the study of reliability. If failure of a component of equipment is completely random (i.e., constitutes an example of a Poisson process), then the lifetimes of these components have an exponential distribution. For a more historical example, one might think of the emergence of organised machine politics in an American city as being a Poisson event, equally likely to occur in any year as in any other, and that the lifetime that a city survives before this happens will have an exponential distribution (Brown and Halaby, 1987).

100. **The normal distribution.** Distributions that have the highest point of the p.d.f. in the middle, with a symmetrical decrease either side, are important in many practical applications. This is roughly the shape of many histograms of data, after all. A particularly important example is the normal distribution,[24] also known as

[24] Following on from footnote 3, here are some more specialist terms we have recently met.

the Gaussian distribution.[25]

- The shape of the p.d.f. is as shown below:

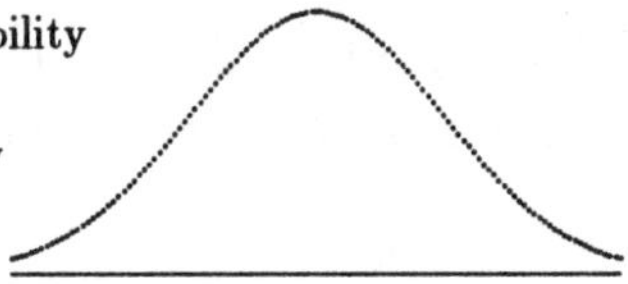

- The symbols μ and σ (Greek letters mu and sigma) are used for the mean and s.d. of the normal distribution, as they are in other contexts when a mean and a standard deviation are known. The mean determines where along the line of real numbers the distribution is located; the standard deviation determines how spread out it is. Some illustrations are in paragraph 106.

- With this distribution, any value of x is possible. However, values outside the range $\mu - 3\sigma$ to $\mu + 3\sigma$ occur only about three times in a thousand — see the final example in paragraph 105 for the calculation of this proportion.

101. **The standard normal distribution.** This has mean $\mu = 0$, and s.d. $\sigma = 1$. It is the usual practice to refer to the horizontal axis as z, rather than x.

- The probability density function is proportional to:

$$\exp\left(-\tfrac{1}{2}z^2\right). \qquad (26)$$

(In case you are not familiar with it, exp(a) is just another way of writing e^a.)

Probability, exclusive, exhaustive, independent, conditional probability, binomial distribution, Poisson distribution, probability density, normal distribution.

It is worthwhile to look them up in a dictionary, and to try to improve the definition you find there.

[25] Named after the German mathematician and astronomer Carl Friedrich Gauss (1777–1855).

Table 1. A (very short) table of the normal distribution.

The column headed $\Phi(z)$ gives the probability that a random variable having a standard normal distribution is less than z.

z	$\Phi(z)$
0.00	0.5000
0.50	0.6915
1.00	0.8413
1.50	0.9332
2.00	0.9772
2.50	0.99379
3.00	0.99865
3.50	0.99977

- As to the cumulative distribution function, there is no elementary formula for this. The symbol Φ (Greek capital letter phi) is used for it, so $\Phi(z)$ is the probability of getting a value less than z when the variable has a normal distribution with mean 0 and s.d. 1.

- The function Φ is not available on most electronic calculators at present, so most statistics textbooks include a table of it. A very short one is printed here as Table 1.

102. **Use of the table of the normal distribution.** Most tables of the normal distribution are arranged so that one chooses z and reads off what Φ is.[26] Thus if we are interested in $z = 1.50$, we read off that Φ is 0.9332. Often, however, the question is slightly different. The reader will find it helpful to draw a little sketch beside each of the following examples, to illustrate what the question is and how it is answered.

[26] With some tables, one chooses z and reads off what $\Phi - 0.5$ is. If you are using one like this, read Table 2.

- What is the probability of getting a value *greater than* a stated value? Required here is $1 - \Phi(z)$. For example, the probability of getting a value greater than 1.50 is 1 − 0.9332, which is 0.0668.

- What is the probability of getting a value *between 0 and* the stated value? Required here is $\Phi(z) - 0.5$ (z being positive).

  Because the standard normal distribution is symmetric about 0, there is a probability of 0.5 to the left of 0, and a probability of 0.5 to the right of 0. To get the probability between 0 and z, we subtract the probability of being less than 0 away from the probability of being less than z.

  For example, the probability of getting a value between 0 and 1.50 is 0.9332 − 0.5000, which is 0.4332.

- Most tables of the normal distribution show only positive values of z. If we want $\Phi(z)$ where z is negative, we have to use the symmetry property in the form $\Phi(-z) = 1 - \Phi(z)$. For example, the probability of getting a value less than −1.50 is $\Phi(-1.50) = 1 - \Phi(1.50) = 1 - 0.9332 = 0.0668$.

- And if we want $1 - \Phi(z)$ where z is negative, we use the symmetry property in the form $1 - \Phi(-z) = \Phi(z)$. For example, the probability of getting a value greater than −1.50 is $\Phi(1.50) = 0.9332$.

- What is the probability of getting a value *between* $-z$ *and* z? By symmetry, this must be twice the probability of getting a value between 0 and z; i.e., $2(\Phi(z)-0.5)$. For example, the probability of getting a value between -1.50 and 1.50 is $2 \times (0.9332 - 0.5000) = 0.8664$.

- What is the probability of getting a value between z_1 and z_2 (where $z_1 < z_2$)? This is $\Phi(z_2) - \Phi(z_1)$. For example, the probability of getting a value between 0.50 and 1.50 is $\Phi(1.50) - \Phi(0.50)$, which is $0.9332 - 0.6915 = 0.2417$. And the probability of getting a value between -0.50 and 1.50 is $\Phi(1.50) - \Phi(-0.50) = \Phi(1.50) - (1 - \Phi(0.50))$, which is $0.9332 - 1 + 0.6915 = 0.6247$.

The student should not attempt to remember the above rules in the form of formulae. Instead, the two key properties that the total probability is 1, and that the distribution is symmetric about 0, should be thoroughly understood.

103. Occasionally, we want the probability of getting a value between z_1 and z_2, where z_1 and z_2 are very close together. In this case, we use the method mentioned in the final sentence of paragraph 96 — we multiply the difference between z_1 and z_2 by the p.d.f. at their average value. The p.d.f. of the normal distribution is:

$$\frac{1}{\sqrt{2\pi}} \exp\left(-\frac{1}{2}z^2\right). \tag{27}$$

For example, to calculate the probability of getting a value of z between 1.500 and 1.501, we multiply $1.501 - 1.500 = 0.001$ by $\frac{1}{\sqrt{2\pi}} \exp\left(-\frac{1}{2}1.5005^2\right)$. And we find $0.001 \times 0.129 = 0.000129$.

104. **Getting a value of z from a given probability.** To do this, we use Table 1 "in reverse". If we require the value of z such that the probability of being less than it equals 0.9772, we see that $z = 2.00$. A more detailed table will enable us to find that

The value of z such that there is a 0.90 probability of being less than it is 1.28,

The value of z such that there is a 0.99 probability of being less than it is 2.33,

etc. Table 1 enables us to directly answer questions about z given the probability (Φ) of being *less* than it, where this probability is 0.5 or greater. Variations from this question require variations in what we do. For example:

- If Φ is less than 0.5, we use the symmetry of the normal distribution to realise that the z that corresponds to Φ is *minus* the z that corresponds to $1 - \Phi$. So if Φ is 0.0228, we look for $1 - 0.0228 = 0.9772$, we find the corresponding z is 2.00, so we know the z that corresponds to 0.0228 is -2.00.

- If we are told the probability of being greater than z, we first need to convert this to Φ before looking in the Table. So if we want the z such that there is a 0.0228 chance of being greater than it, we look for $\Phi = 0.9772$, and find that z is 2.00.

- For what value of z does the probability of being within z of 0 equal 0.95? By symmetry, the probability 0.05 of being further away than z must be equally divided, 0.025 being the the probability of being less than $-z$ (and 0.025 also being the probability of being greater than z). Hence what we want is the value of z such that Φ is $0.025 + 0.95 = 0.975$. We find (from a detailed table) that $z = 1.96$.

Table 2. Read this if your table of the normal distribution gives $\Phi - 0.5$.

Some tables of the normal distribution do not give the probability of getting a value less than z. Instead, they give the probability of getting a value between 0 and z. That is, when you know z, you read off what $\Phi - 0.5$ is. A very short example of such a table is as follows:

z	$\Phi(z) - 0.5$
0.00	0.0000
0.50	0.1915
1.00	0.3413
1.50	0.4332
2.00	0.4772
2.50	0.49379
3.00	0.49865
3.50	0.49977

Some examples of the use of such a table are now given. The reader will find it helpful to draw a little sketch beside each of the examples, to illustrate what the question is and how it is answered.

- What is the probability of getting a value between 0 and 1.50? Directly from the table, this is 0.4332.
- What is the probability of getting a value less than 1.50? Here, we need to add on 0.5 to the value shown in the table, and we get 0.9332. (Because the standard normal distribution is symmetric about 0, there is a probability of 0.5 to the left of 0, and a probability of 0.5 to the right of 0. To get the probability between $-\infty$ and z, we add the probability of being less than 0 to the probability of being between 0 and z.)
- What is the probability of getting a value between -1.50 and 1.50? By...

Table 2 (continued).

... symmetry, this must be twice the probability of getting a value between 0 and 1.50; that is, $2 \times 0.4332 = 0.8664$.

- What is the probability of getting a value greater than 1.50? We subtract the probability of being between 0 and 1.50 from the probability of being greater than 0 (which we know to be 0.5000); so the answer is $0.5000 - 0.4332 = 0.0668$.
- What is the probability of getting a value less than −1.50? By symmetry, this must be the same as the probability of getting a value greater than 1.50; that is, 0.0668.

Here are some examples of going from a probability to a value of z:

- Given that the probability of being between 0 and z is 0.4772, what is the value of z? Using the table "in reverse", we see this is 2.00.
- If the probability of being less than z is 0.9772, what is the value of z? We first subtract 0.5000 from 0.9772 and get 0.4772, and look for this when using the table in reverse.
- For what value of z does the probability of not being within z of 0 equal 0.05? By symmetry, the probability 0.05 must be equally divided, 0.025 being the probability of being greater than z (and 0.025 also being the probability of being less than $-z$). Hence what we want is the value of z such that $\Phi - 0.5$ is 0.475. We find (from a detailed table) that $z = 1.96$.

- For the normal distribution, what is the ratio of the inter-quartile range to the s.d.? The upper quartile of the normal distribution is the value of z such that $\Phi(z) = 0.75$. From a detailed table of the normal distribution, we find $z = 0.67$. By symmetry, the lower quartile is -0.67. The IQR is therefore 1.34. Since the s.d. is 1, the ratio of the IQR to the s.d. is 1.34.

- For the normal distribution, what proportion of observations are outliers, if we adopt the definition of "outlier" given in paragraph 22? The upper quartile plus 1 IQR is $0.67 + 1.34 = 2.01$. From a detailed table of the normal distribution, we find that a proportion 0.0444 of observations are either bigger than 2.01 or smaller than -2.01.

105. **The normal distribution with some other mean and standard deviation.** So far, we have been dealing with the *standard* normal distribution — that is, the one with mean = 0 and s.d. = 1. Now, what do we do in the practical case where the mean and s.d. are something else? There is a simple answer: we can transform questions about a value x from the normal distribution with mean μ and s.d. σ into questions about a value z from the standard normal distribution, by calculating

$$z = \frac{x - \mu}{\sigma}. \tag{28}$$

The reader will again find it helpful to draw little sketches of what the question is (with x, μ, and σ shown), and what the equivalent question concerning the standard normal distribution is (with z shown).

- If a random variable has a normal distribution with mean 10 and s.d. 5, what is the probability it will take a value less than 12.5?

In this question, x is 12.5, μ is 10, and σ is 5. To answer it, we calculate $z = \frac{12.5-10}{5} = 0.50$. Then we look in Table 1 and find that Φ is 0.6915.

- Suppose that women's heights are normally distributed with a mean of 1624 mm. and a s.d. of 56 mm. What proportion of women are shorter than 1500 mm.?

  In this question, x is 1500, μ is 1624, and σ is 56. To answer it, we calculate $z = \frac{1500-1624}{56} = -2.21$. The probability of getting a z less than -2.21 equals the probability of getting a z greater than 2.21 (using the symmetry property), which equals 1 minus the probability of getting a z less than 2.21. From a detailed table of the normal distribution, we find this is 0.9864, so 0.0136 is the required answer.[27]

- Suppose that women's heights are normally distributed with a mean of 1624 mm. and a s.d. of 56 mm. What height is such

[27] Some instructors like you to set the calculation out as below.

$$\begin{aligned}\Pr\{X < 1500\} &= \Pr\left\{\frac{X-1624}{56} < \frac{1500-1624}{56}\right\} \text{ (where we are doing the same thing to the right hand side of the < sign as we are to the left)}\\ &= \Pr\{Z < -2.21\}\\ &= 0.0136.\end{aligned}$$

that 5 per cent of women are shorter than it?[28]

To answer this sort of question, we realise that we can

Go from a probability to a value of z using the table of the normal distribution, and

Go from a value of z to a value of x using the formula $z = \frac{x-\mu}{\sigma}$.

So, from $\Phi = 0.05$ we find $z = -1.645$. Then, knowing $-1.645 = \frac{x-1624}{56}$, we obtain $x = (-1.645 \times 56) + 1624 = 1532$.

- What proportion of observations are further than 3 standard deviations away from the mean?

  In this question we are asking what proportion of observations are less than $\mu - 3\sigma$ or are greater than $\mu + 3\sigma$. Do not worry that we are not given the x's as numerical values, we will see that the unknown quantities cancel out. Putting $x_1 = \mu - 3\sigma$, we find $z_1 = \frac{\mu-3\sigma-\mu}{\sigma} = -3$; putting $x_2 = \mu + 3\sigma$, we find $z_2 = \frac{\mu+3\sigma-\mu}{\sigma} = 3$. From Table 1, we find the probability is $0.00135 + 0.00135 = 0.0027$.

105x. Perhaps you think that the heights of people is a subject no historian is interested in? How wrong you are! Two examples of recent research are as follows.

- Komlos (1993) argued that the construction of conventional indices of standard of living, based upon economic variables such as hourly wages, is "fraught with numerous uncertainties". As an alternative, "Human stature can be used a proxy measure of the biological well-being of a population, an important component of that complex variable called the standard of living."

[28] As mentioned in footnote 10, vehicles are typically designed to be safe for the 5th percentile female and the 95th percentile man, as well as for people of average size. If one has measured a large number of people, one can read off from the cumulative frequency plot what the 5th percentile is. But if one has only quite a small sample of people, one might prefer to calculate the mean and s.d., assume the normal distribution is valid here, and derive the 5th percentile from that, as we are now doing.

However, the conclusions to be drawn from available data are controversial. Whereas Floud et al (1993) found an overall upward trend in British heights in the later 18th and early 19th centuries, punctuated by a temporary downturn, Komlos (1993) found an overall downward trend.

- A study of slaves shipped from the east coast of the U.S.A. to New Orleans concluded that the net nutritional status of U.S. slaves was better than that of most European workers during the 19th century, but that the slave children were extremely short, presumably due to malnourishment (Pritchett and Freudenberger, 1992).

106. **Expression for the p.d.f. of the normal distribution, and illustrations.** The p.d.f. is:

$$\frac{1}{\sigma\sqrt{2\pi}} \exp\left[-\frac{1}{2}\left(\frac{x-\mu}{\sigma}\right)^2\right]. \tag{29}$$

The two distributions shown here differ in μ but have the same σ:

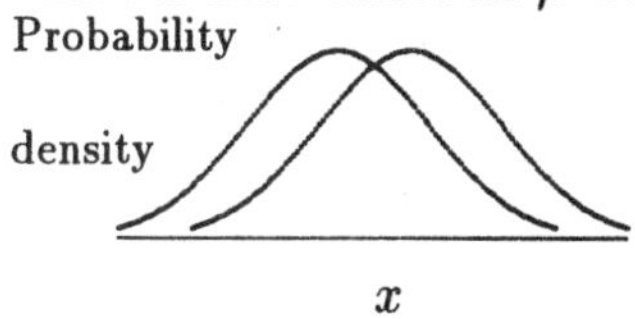

These two distributions differ in σ but not in μ:

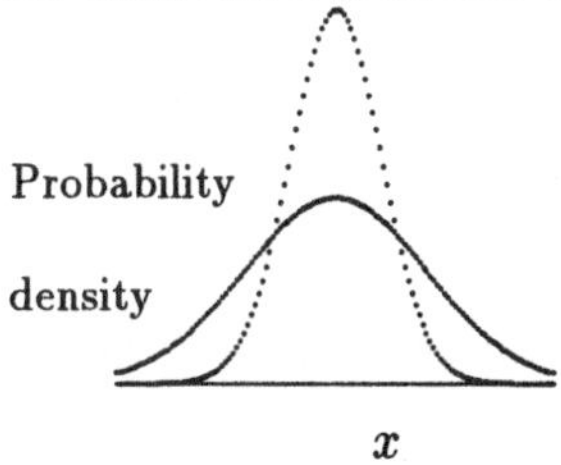

These two distributions differ in both μ and σ:

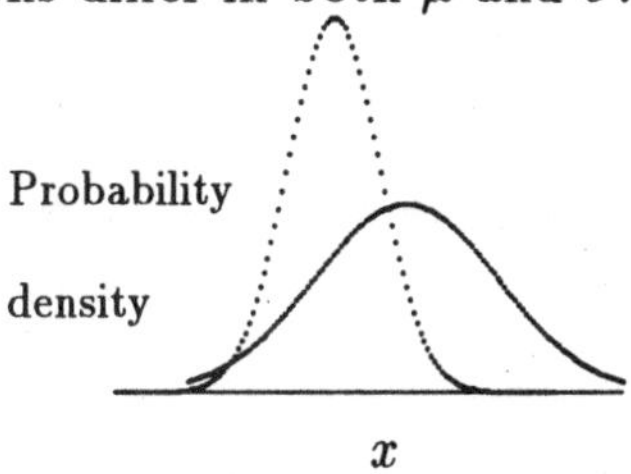

107. **Using the cumulative frequency plot to graphically judge whether data is normally distributed.** A cumulative frequency plot (paragraphs 27 and 36) would be easier to comprehend if it was (at least approximately) a straight line. Shown below are the cumulative proportions for the arm length data of paragraph 32. (These have been repeated from paragraph 36.) If these proportions referred to a normal distribution, they would be Φ's, and could be converted into z's by using a detailed table of the normal distribution in reverse, as in paragraph 104. This has been done in the final column.

up to 699	.01	−2.33
up to 719	.03	−1.88
up to 739	.09	−1.34
up to 759	.22	−0.77
up to 769	.34	−0.41
up to 779	.42	−0.20
up to 789	.53	0.08
up to 799	.60	0.25
up to 809	.76	0.71
up to 829	.87	1.13
up to 849	.98	2.05

If we plot the z's that have been obtained in this way against the arm lengths, a straight line will indicate that arm length is normally distributed.[29]

[29] The reason is as follows. To say that arm length is normally distributed is to say that

$$\Pr\{\text{Arm length} < x\} = \Phi\left(\frac{x-\mu}{\sigma}\right),$$

for some mean μ and s.d. σ. Now, we know that if we start with a value ζ, convert it to $\Phi(\zeta)$, and take the z corresponding to this Φ, we get back to ζ; that is, $z(\Phi(\zeta)) = \zeta$; expressed in other words, z is the inverse transformation of Φ. So, taking the z corresponding to both sides of the equation,

$$\begin{aligned} z(\Pr\{\text{Arm length} < x\}) &= z\left(\Phi\left(\tfrac{x-\mu}{\sigma}\right)\right) \\ &= \frac{x-\mu}{\sigma}, \end{aligned}$$

a relation that is linear in x.

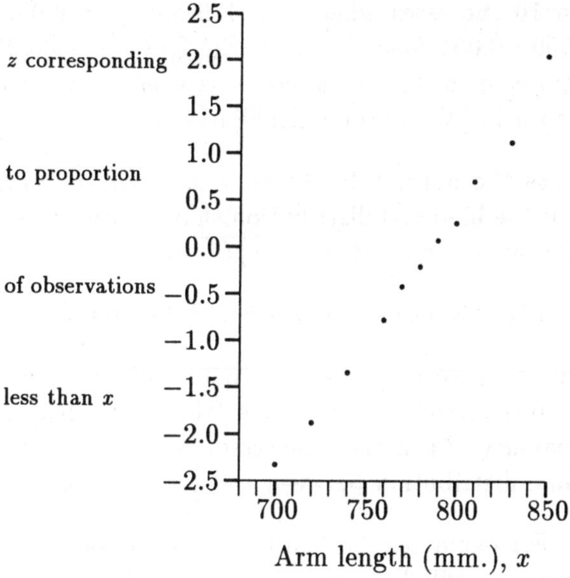

Remarks:

- As in paragraph 36, the points have been plotted at arm lengths of 700, 720, etc., rather than 699, 719, etc.

- It is clear that a normal distribution is a good overall description of this data, despite the irregularities that are evident in the histogram in paragraph 35.

- Special graph paper is available to help with the procedure described: instead of converting the cumulative probabilities into z's, the probabilities are plotted directly onto the vertical scale of the graph paper — which is not a linear scale like ordinary graph paper is, but is non-linear in just the way required for this procedure to work.

- If the values of the observations are known exactly, rather than being grouped into classes, we treat $(i - \frac{1}{2})/n$ as being the appropriate value of Φ for the ith smallest observation, convert these Φ's into z's, and plot against the corresponding value of the observation, x.

108. **Using the normal distribution to approximate the binomial.** Suppose we want to work out the probability of not

more than 10 successes when $n = 150$ and $p = 0.05$. We require $0.95^{150} + 150 \times 0.05 \times 0.95^{149} + \ldots + \frac{150 \times 149 \times \ldots \times 141}{10 \times 9 \times \ldots \times 1} \times 0.05^{10} \times 0.95^{140}$, which is quite tedious to work out. A reasonable approximation can be obtained using the normal distribution.

- We use the normal distribution that has the same mean and s.d. as the binomial distribution that we are interested in. That is, we set $\mu = np$ and $\sigma = \sqrt{np(1-p)}$.

  These work out to be 7.5 and 2.67 for this example.

- Notice that we are using a continuous distribution (the normal) to approximate a discrete one (the binomial). The binomial probability of exactly x successes will be approximated by the normal distribution between $x - \frac{1}{2}$ and $x + \frac{1}{2}$.

  For example, the binomial probability of exactly 10 successes will be approximated by the normal distribution between 9.5 and 10.5, the binomial probability of exactly 9 successes will be approximated by the normal distribution between 8.5 and 9.5, and so on.[30] So we can see that the binomial probability of 10 or fewer successes will be approximated by the normal distribution over the range extending up as far as 10.5.

- Now we know what distribution we need to use (the normal with mean 7.5 and s.d. 2.67), and what area we want (to the left of 10.5). We calculate $z = \frac{10.5 - 7.5}{2.67} = 1.12$. From a table of the normal distribution, we find the required probability is 0.8686.

- This method of approximation is not satisfactory if n is small or p is very different from 0.5. A common recommendation is to only use it if $np > 5$ and $n(1-p) > 5$.

The mathematics that connects the binomial probability (18) to the normal p.d.f. (29) is too advanced for us — but it does exist!

[30] The addition or subtraction, as appropriate, of 0.5 is known as the *continuity correction.*

109. Questions on approximating the binomial distribution by the normal are quite popular (with examiners!) because they enable the testing of several ideas at once — recognising the general strategy that is required, knowing how to calculate the mean and s.d. of the binomial distribution, appreciating that the limit of integration for the normal distribution will differ by 0.5 from the x for the binomial distribution, the conversion into z, and the use of the table of the normal distribution.

> A common cause of error is adding 0.5 when subtracting is required, and *vice versa.* To avoid this, think carefully about what range of binomial outcomes is referred to in the question; if this is between whole numbers x_1 and x_2 inclusive (with $x_1 < x_2$), the normal approximation will refer to the range $x_1 - 0.5$ to $x_2 + 0.5$. (But it is usual to use $-\infty$ if x_1 is 0, and ∞ if x_2 is n.)

110. **Expectations of discrete random variables.** As before, let the probability that a random variable X takes the value x be $\Pr\{X = x\}$. Then the *expectation* of X, written $E(X)$, is

$$\Sigma\, x \,.\, \Pr\{X = x\}, \tag{30}$$

with the summation taking place over all the possible values of X. That is, the expectation is in effect the mean. Suppose, for example, X has the following distribution:

x:	0	1	2	3	4
$\Pr\{X = x\}$:	0.1	0.2	0.4	0.2	0.1

Then

$$\begin{aligned} E(X) &= (0 \times 0.1) + (1 \times 0.2) + (2 \times 0.4) + (3 \times 0.2) \\ &\quad + (4 \times 0.1) \\ &= 0 + 0.2 + 0.8 + 0.6 + 0.4, \end{aligned}$$

which totals 2.0. (This answer is actually obvious, because of the symmetry of the distribution.)

111. Notice that the "expected" value is not necessarily the most frequent value — nor is it necessarily even a possible value! For example, if X takes the values -1 and 1 with probabilities of $\frac{1}{2}$ each, then $E(X)$ is 0. So, "expected" value is something of a misnomer.

112. **Expectations of continuous random variables.** In this case, integration replaces summation and $E(X)$ is

$$\int x f(x)\, dx, \tag{31}$$

where $f(x)$ is the p.d.f. Suppose, for example, $f(x)$ is $1 - \frac{1}{2}x$ for x being between 0 and 2, and is 0 for other x. Then

$$E(X) = \int_0^2 x\left(1 - \tfrac{1}{2}x\right)\, dx = \left[\tfrac{1}{2}x^2 - \tfrac{1}{6}x^3\right]_{x=0}^{x=2},$$

which equals $\frac{2}{3}$. Integration in (31) (and (33) below) is from $-\infty$ to ∞; but the example of $f(x)$ that we are considering is 0 outside the range $x = 0$ to $x = 2$.

113. **Variances and standard deviations of discrete random variables.** The variance of a random variable, $V(X)$, is the expectation of $[X - E(X)]^2$, that is,

$$\Sigma\,[x - E(X)]^2 . \Pr\{X = x\}. \tag{32}$$

Thus for the example of paragraph 110,

$$\begin{aligned} V(X) &= [(-2)^2 \times 0.1] + [(-1)^2 \times 0.2] + [0^2 \times 0.4] \\ &\quad + [1^2 \times 0.2] + [2^2 \times 0.1] \\ &= 0.4 + 0.2 + 0 + 0.2 + 0.4, \end{aligned}$$

which totals 1.2. The standard deviation is the square root of the variance, so in this example it is 1.10.

114. **Variances and standard deviations of continuous random variables.** Again, integration replaces summation and $V(X)$ is

$$\int [x - E(X)]^2 f(x)\, dx. \tag{33}$$

Thus for the example of paragraph 112,

$$\begin{aligned} V(X) &= \int_0^2 \left(x - \tfrac{2}{3}\right)^2 \left(1 - \tfrac{1}{2}x\right)\, dx \\ &= \left[-\tfrac{1}{8}x^4 + \tfrac{5}{9}x^3 - \tfrac{7}{9}x^2 + \tfrac{4}{9}x\right]_{x=0}^{x=2}, \end{aligned}$$

which equals $\frac{2}{9}$. (And the standard deviation is $\sqrt{\frac{2}{9}} = 0.47$.)

115. **A more convenient formula for the variance.** In the context of data analysis (paragraphs 15–17), $\frac{1}{n-1}[\Sigma\, x^2 - \frac{1}{n}(\Sigma\, x)^2]$ is a more convenient formula than $\frac{1}{n-1}\Sigma\,(x - \bar{x})^2$ for computing the variance. Similarly,

$$E(X^2) - [E(X)]^2 \qquad (34)$$

is here the most convenient formula.[31] That is, the expectation of X^2 minus the square of the expectation of X.

- For our discrete example, we have $E(X^2) = (0 \times 0.1) + (1 \times 0.2) + (4 \times 0.4) + (9 \times 0.2) + (16 \times 0.1) = 0 + 0.2 + 1.6 + 1.8 + 1.6$, which totals 5.2. Thus $V(X) = 5.2 - 2^2 = 1.2$, which is the result we calculated previously.

- Another discrete example is given in Table 3.

- For our continuous example, $E(X^2) = \int_0^2 x^2(1 - \frac{1}{2}x)\, dx = \left[\frac{1}{3}x^3 - \frac{1}{8}x^4\right]_{x=0}^{x=2}$, which is $\frac{2}{3}$. Thus $V(X) = \frac{2}{3} - \left(\frac{2}{3}\right)^2 = \frac{2}{9}$, as previously.

- Now consider the exponential distribution (paragraph 99). This has p.d.f. $f(x) = \frac{1}{\theta}e^{-x/\theta}$ (provided x is positive). The expectation $E(X) = \int_0^\infty x.\frac{1}{\theta}e^{-x/\theta}\, dx$ can be found to be θ (using the method of integrating by parts). The expectation of the

[31] To show that this is the expectation of $[X - E(X)]^2$, we simply write

$$\begin{aligned} E([X - E(X)]^2) &= E(X^2 - 2X\,E(X) + [E(X)]^2) \\ &= E(X^2) - 2E(X).E(X) + [E(X)]^2 \\ &\quad \text{(since } E(X) \text{ is a constant)} \\ &= E(X^2) - [E(X)]^2. \end{aligned}$$

Table 3. The expectation and s.d. of the day of the month on which people are born.

Let us assume that the same number of people are born on each day, regardless of which month or year it is. Because of leap years, we need to consider a 4-year period. In such a period, there are 28 months with 31 days, 16 months with 30 days, 1 month with 29 days, and 3 months with 28 days. There are 1461 days altogether. Consequently, the proportions of people born on the respective days of the month will be as below:

Day of the month, x:	1	2	3	...	28	29	30	31
$\Pr\{X = x\}$:	$\frac{48}{1461}$	$\frac{48}{1461}$	$\frac{48}{1461}$	...	$\frac{48}{1461}$	$\frac{45}{1461}$	$\frac{44}{1461}$	$\frac{28}{1461}$

- To compute $E(X)$, we require $\Sigma_{x=1}^{28} x.\frac{48}{1461} + 29.\frac{45}{1461} + 30.\frac{44}{1461} + 31.\frac{28}{1461}$. The first term is $\frac{48}{1461}\Sigma_{x=1}^{28} x$. Most students know from their mathematics course that $\Sigma_{x=1}^{m} x = \frac{1}{2}m(m+1)$. Therefore, the first term is $\frac{48}{1461} \times \frac{1}{2} \times 28 \times 29$, which is $\frac{19488}{1461}$. Adding on the other terms gives a total of $\frac{22981}{1461} = 15.73$.

- To compute $E(X^2)$, we require $\Sigma_{x=1}^{28} x^2.\frac{48}{1461} + 29^2.\frac{45}{1461} + 30^2.\frac{44}{1461} + 31^2.\frac{28}{1461}$. The first term is $\frac{48}{1461}\Sigma_{x=1}^{28} x^2$. Most students know from their mathematics course that $\Sigma_{x=1}^{m} x^2 = \frac{1}{6}m(m+1)(2m+1)$. Therefore, the first term is $\frac{48}{1461} \times \frac{1}{6} \times 28 \times 29 \times 57$, which is $\frac{370272}{1461}$. Adding on the other terms gives a total of $\frac{474625}{1461} = 324.86$.

- Finally, $V(X) = E(X^2) - [E(X)]^2 = 77.44$, and the s.d. is $\sqrt{77.44} = 8.80$.

This value of the s.d. will be used in paragraph 133.

square $E(X^2) = \int_0^\infty x^2.\frac{1}{\theta}e^{-x/\theta}\,dx$ can be found to be $2\theta^2$ (by the same method). Hence the variance $V(X) = 2\theta^2 - \theta^2 = \theta^2$, and the s.d. is θ.

116. **Waiting time.** If X represents the time gap between successive buses, and people are coming to the bus stop at a constant rate, then their average waiting time is $\frac{1}{2}E(X^2)/E(X)$ (end of paragraph 10). Now, $\frac{1}{2}E(X^2)/E(X) = \frac{1}{2}\{V(X) + [E(X)]^2\}/E(X)$ $= \frac{1}{2}(\sigma^2 + \mu^2)/\mu$ (where μ and σ are the mean and s.d. of the gaps between buses), and this may be written as $\frac{1}{2}\mu\left(1 + \frac{\sigma^2}{\mu^2}\right)$, which is the form that this formula is usually seen in.

If the distribution of gap lengths is exponential with mean θ, the average waiting time is $\frac{1}{2}\theta\left(1 + \frac{\theta^2}{\theta^2}\right) = \theta$, twice as much as it would be if the buses came regularly (i.e., with $\sigma = 0$).

117. **Effect on the expectation and variance of adding a constant.** If we add a constant, a, on to each value of a random variable, the effect on the expectation is that a is added to it. There is no change to the variance or standard deviation. In symbols,

$$E(a+X) = a + E(X) \qquad (35)$$
$$V(a+X) = V(X). \qquad (36)$$

118. **Effect on the expectation and variance of multiplying by a constant.** If we multiply each value of a random variable by a constant, b, the effect on the expectation is that it is multiplied by b. The standard deviation is multiplied by b, and the variance is multiplied by b^2. In symbols,

$$E(b.X) = b.E(X) \qquad (37)$$
$$V(b.X) = b^2.V(X). \qquad (38)$$

119. For example, suppose the numbers 20, 25, 30 are temperatures in degrees Celsius. We multiply by $\frac{9}{5}$ and add 32 to convert them to degrees Fahrenheit: 68, 77, 86. The same thing has happened to the mean — it has been multiplied by $\frac{9}{5}$ and then 32 added.

The effect on how spread out the numbers are has been to multiply by $\frac{9}{5}$. So the s.d. has been multiplied by $\frac{9}{5}$ (and hence the variance has been multiplied by $\left(\frac{9}{5}\right)^2$).

120. The expectation when multiplying together two independent random variables. If X_1 and X_2 are two *independent* random variables, the expectation of their product is the product of the expectations,[32]

$$E(X_1X_2) = E(X_1)E(X_2). \tag{39}$$

121. The expectation and variance of a sum of random variables. Suppose we have two random variables, X_1 and X_2, and we create a new random variable by adding them together, $X_1 + X_2$. The expectation of the sum is the sum of the expectations,

$$E(X_1 + X_2) = E(X_1) + E(X_2). \tag{40}$$

Furthermore, *provided X_1 and X_2 are independent*, the variance of the sum is the sum of the variances,[33]

$$V(X_1 + X_2) = V(X_1) + V(X_2). \tag{41}$$

---

[32] The crucial step in proving this is to recall that the probability of a particular combination of values is the product of the individual probabilities, provided that independence holds.

[33] To prove this, write

$$\begin{aligned} E((X_1+X_2)^2) &= E(X_1^2 + 2X_1X_2 + X_2^2) \\ &= E(X_1^2) + 2E(X_1X_2) + E(X_2^2) \\ &= E(X_1^2) + 2E(X_1)E(X_2) + E(X_2^2) \\ &\quad \text{(because } X_1 \text{ and } X_2 \text{ are independent).} \\ [E(X_1+X_2)]^2 &= [E(X_1) + E(X_2)]^2 \\ &= [E(X_1)]^2 + 2E(X_1)E(X_2) + [E(X_2)]^2. \end{aligned}$$

Subtracting the two equations, the left hand side is $E((X_1+X_2)^2) - [E(X_1+X_2)]^2$, which is $V(X_1+X_2)$; the right hand side is $E(X_1^2) - [E(X_1)]^2 + E(X_2^2) - [E(X_2)]^2$, which is $V(X_1) + V(X_2)$.

Actually, this shows that equation (41) holds if $E(X_1X_2) = E(X_1)E(X_2)$. If this holds, X_1 and X_2 are said to be *uncorrelated*. X_1 and X_2 are uncorrelated if they are independent; but they might possibly be uncorrelated even if they are not independent.

Naturally, we need not stop at two variables, we can add together as many as we like. And the expectation of a difference is the difference of the expectations,

$$E(X_1 - X_2) = E(X_1) - E(X_2), \qquad (42)$$

and the variance of a difference between *independent* random variables is the sum of the variances,

$$V(X_1 - X_2) = V(X_1) + V(X_2). \qquad (43)$$

The variance of a difference is the *sum* of the variances, not the difference. This is common sense — after all, there are two sources of variation contributing to the variation in $X_1 - X_2$, the variation of X_1 and the variation of X_2; obviously, the total variation is going to be larger than either individual source. Furthermore, we are adding X_1 and $-1 \times X_2$, and we know that the variance of $b \times X_2$ is $b^2 \times$ the variance of X_2; here b is -1 and b^2 is 1.

The standard deviation of a sum or difference of independent random variables should be calculated using equation (41) or (43), together with the fact that the variance is the square of the s.d.

122. As an example, suppose a pen is made up of a barrel (length B) that fits into a cap (length C); the one fits over the other by an amount R. The pen has to fit into a wallet (length W). Knowing the expectations and variances of B, C, R, and W, determine the expectation and s.d. of the difference in lengths between the wallet and the pen.

The length of the pen is $B + C - R$. The amount by which the wallet length exceeds the pen length is $W - B - C + R$. Let us suppose the expectations and s.d.'s of the measurements are as shown below (in cm.).

	W	B	C	R
Expectation:	14.0	13.0	5.0	4.4
s.d.:	0.1	0.1	0.05	0.2

Then $E(W - B - C + R) = E(W) - E(B) - E(C) + E(R) = 14.0 - 13.0 - 5.0 + 4.4$, which is 0.4. And $V(W - B - C + R) = V(W) + V(B) + V(C) + V(R) = 0.1^2 + 0.1^2 + 0.05^2 + 0.2^2$, which is 0.0625, and so the s.d. is 0.25.

If we could assume a normal distribution for the difference in lengths between the wallet and the pen, we could go on to determine the probability of the pen being able to fit into the wallet, that is, the probability of $W - B - C + R$ exceeding 0.

123. As a further example, suppose that the yield of metal from some primitive smelting process has a mean of 15 ounces and a s.d. of 4 ounces. If 8 batches of the raw materials are smelted, what is the expectation and s.d. of the total yield? Writing Y for the total yield, we have:

	Y	=	X_1	+	X_2	+	...	+	X_8
Expectation:			15		15		...		15
Variance:			4^2		4^2		...		4^2

So the expectation of the total is $8 \times 15 = 120$, and the variance is $8 \times 16 = 128$ (and the s.d. is $\sqrt{128} = 11.31$).

124. **The variance of the binomial distribution.** In paragraph 85, the variance of the binomial distribution was stated to be $np(1 - p)$.

- One way of proving this is directly from $E(X^2) = \Sigma\, x^2.\Pr\{X = x\} = \Sigma\, x^2 \binom{n}{x} p^x (1 - p)^{n-x}$. But to proceed further requires some facility in manipulating combinatorial expressions.

- I think it is easier to work out the variance as follows.

  - When $n = 1$, the outcome 1 is observed with probability p, and the outcome 0 is observed with probability $1-p$. The variance is $E(X^2) - [E(X)]^2$. We find $E(X^2) = 0^2.(1 - p) + 1^2.p = p$. And $E(X) = 0.(1 - p) + 1.p = p$. So the variance is $p - p^2 = p(1 - p)$.
  - Now realise that the number of successes in n trials is like adding up $X_1 + X_2 + \ldots + X_n$, where X_i is the number of successes (0 or 1) in trial i.

- As the variance of the sum of independent random variables is the sum of the variances, the variance of the number of successes in n trials is $np(1-p)$.

125. **Do not confuse $X_1+X_2+\ldots+X_n$ with nX_1.** Suppose that $X_1, X_2, \ldots X_n$ are all independent of each other, and that they have identical distributions. Since the distributions are identical, the expectations are all identical and the variances are all identical, and we might as well refer to each of them as $E(X_1)$ and $V(X_1)$, respectively. When we add all the X_i's together, the expectation and variance of the total are $nE(X_1)$ and $nV(X_1)$. Be careful not to confuse this situation with the one in which we have a *single* random variable X_1, and are multiplying it by n. In this latter case, the expectation and variance are $nE(X_1)$ and $n^2V(X_1)$ (see paragraph 118).

126. **The sum of normally-distributed random variables.** If random variables X_1 and X_2 each have normal distributions, then their sum X_1+X_2 also has a normal distribution. (We will not prove this.) So, therefore, does their difference, and this will be used in paragraphs 170–171 when performing a hypothesis test on the difference between two means.

Part III. Inference

127. **Introduction.** The aim in this Part is to use the tools of probability, from Part II, to extend the methods of data description outlined in Part I. We will try to draw conclusions about the populations from which our random samples came — that is, to make inferences. Paragraphs 144 onwards will deal with *confidence intervals* and *hypothesis testing*, and will include discussion of the meaning of what is being done, as well as instruction in how to do it. As foundation for these topics, paragraphs 128–143 are given over to the concept of *standard error*, and the *Central Limit Theorem.*

128. **The mean in different samples.** Early in a statistics course, one learns how to calculate the mean of a sample. However, if another sample is drawn and its mean calculated, there will be a slightly different answer. A third sample will give a different answer again. Now, if we could describe the sample-to-sample variation, we could draw conclusions concerning which possible values for the population mean were consistent with the data, and which were not.

> What you need to get used to now is the idea that the sample mean is at the same time both
>
> > One specific number (for this specific sample), calculated by a particular formula, and
> >
> > A random variable which will vary from sample to sample.

129. **Sampling distributions.** The mean is an important statistic that summarises one aspect of data. The particular value that is obtained will be different in different samples. Thus it is said to have a sampling distribution. Similarly, other statistics (e.g., the

median, and the s.d.) will also differ from sample to sample, and will have their own sampling distributions.

130. **The standard error of the mean.** The amount of variability in a quantity can be measured by its standard deviation. So the amount of variability in the mean is measured by the standard deviation of the mean — the standard deviation, that is, calculated over repeated taking of samples.

> Sample A consists of n observations from a very large population. These observations are $a_1, a_2, \ldots a_n$. They have mean $\bar{a}$. Taking another sample of n observations, B, we find observations $b_1, b_2, \ldots b_n$, which have mean $\bar{b}$. Sample C has another n observations, the mean of which is $\bar{c}$. Sample D has another n observations, the mean of which is $\bar{d}$. And so on. The standard deviation of the mean is the standard deviation of the numbers $\bar{a}, \bar{b}, \bar{c}, \bar{d}, \ldots$

To try to avoid confusion with the standard deviation that applies to individual observations, the standard deviation of the mean is given a special name — the *standard error of the mean.*

131. **Formula for the standard error of the mean.** We usually do not have several samples from the same population. So we cannot find the standard error of the mean in the way described above. Instead, the following equation connects the variability in the mean to the variability that applies to an individual observation:[34]

$$\sigma_{\bar{x}} = \frac{\sigma}{\sqrt{n}}, \tag{44}$$

[34] This formula may be derived from what we know about the variance of a sum of independent random variables.

- If $X_1, X_2, \ldots X_n$ are n independent observations from a population that has a mean μ and s.d. σ, then the variance of the total $T = (X_1 + X_2 + \ldots + X_n)$ is $n\sigma^2$ (see paragraphs 121 and 125).
- The variance of $\frac{1}{n}T$ must be $\frac{1}{n^2} \times$ the variance of T (see paragraph 118). So the variance of $\frac{1}{n}T$ is $\frac{1}{n^2}n\sigma^2 = \frac{1}{n}\sigma^2$.
- And the s.d. of $\frac{1}{n}T$ is $\sigma/\sqrt{n}$.
- Of course, $\frac{1}{n}T$ is the sample mean $\bar{x}$.

where $\sigma_{\bar{x}}$ is the standard error of the mean, and σ is the s.d. that applies to individual observations.

132. **Interpretation of the standard error of the mean.** It measures the extent of random variation in sample means.

- If $\sigma_{\bar{x}}$ is small, the chances are good that the mean of a sample, $\bar{x}$, will be close to the true mean of the population, μ.
- If $\sigma_{\bar{x}}$ is large, we are more likely to get a sample mean that differs considerably from the mean of the population.

Two factors determine $\sigma_{\bar{x}}$:

- How variable the observations are, σ.
- How big a sample we take.

In order to have an accurate estimate of μ, we require that σ be small, and/or that n be large.

- The variability of the observations, σ, may be largely due to the variability in the items being observed; or there may be appreciable measurement error, also. There is usually nothing that can be done about the inherent variability in the items, but it may be possible to use a more accurate method of measurement (perhaps a more expensive one).
- The other strategy for making $\sigma_{\bar{x}}$ small is to take a large sample; the disadvantage of this is the extra expense. Notice that the square root of n is what is in the equation — this means that in order to halve $\sigma_{\bar{x}}$, for example, we need to quadruple n.

133. The following exercise demonstrates what the standard error of the mean really signifies, and that it is smaller for larger samples. To make the description clear, I will assume the class has 117 students in it, but the exercise works just as well with any other reasonably large number.

1. For each student, let their value of X be the day of the month on which they were born. (Thus we know each value of X is

between 1 and 31, inclusive.) The task is to estimate μ_X, the average day of the month on which students were born. (It is obvious that this is approximately 16; in Table 3, we found that it is 15.73.)

2. Let the class be divided into 3 groups of 25 students each, 4 groups of 9 students each, and 6 groups of 1 student each.

3. Each student tells the others in their group their birthday.

4. Each group works out its average birthday. A spokesperson announces this to the whole class. (In the calculation below, I will assume these averages have turned out to be 15, 19, 16, 17, 19, 10, 15, 9, 2, 21, 30, 17, 14.)

5. The results can be summarised in a table as below.

Group no.	How many?	Its mean
1	25	15
2	25	19
3	25	16
4	9	17
5	9	19
6	9	10
7	9	15
8	1	9
9	1	2
10	1	21
11	1	30
12	1	17
13	1	14

6. We can now work out the standard deviation of the means based upon samples of size 25. And that for means based upon samples of size 9. And that for means based upon samples of size 1. For the above data, the results are as follows:

Sample size	s.d.
25	2.1
9	3.9
1	9.7

7. Now the s.d.'s can be plotted against n, and it can be seen that the larger n is, the smaller the s.d. is. (Taking into account the differing lengths of months and assuming equal numbers of people are born on each day, we found in Table 3 that σ is 8.80; so the theoretical values for the above numbers are $\frac{8.80}{\sqrt{25}} = 1.8$, $\frac{8.80}{\sqrt{9}} = 2.9$, and $\frac{8.80}{\sqrt{1}} = 8.8$.)

133x. **Understanding what someone else has drawn.** In reports of research, you will sometimes come across pictures like this:

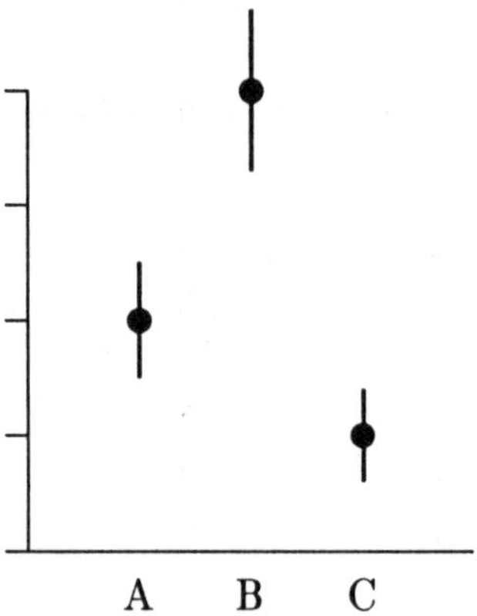

Do not let your eyes skim lightly across an illustration like this. Make sure you understand what it is showing. The locations of the dots show what the means of the three groups A, B, C, were, obviously enough, but what do the bands either side of the means represent? You can only answer this by reading the caption to the picture and the text that accompanies it. The two main questions that should be in your head when doing this are:

- Standard deviation or standard error?
- One or two?

That is, are the bands illustrating the variability in the observations (the standard deviations), or are they illustrating how accurately the means have been estimated (the standard errors)? And do they extend ±1, or ±2, or perhaps some other number, of standard deviations (or standard errors) from the means?

134. **Standard errors of the median and s.d.** How accurately a median or a standard deviation has been estimated is not usually studied in an introductory statistics course, but students are often interested to know approximately the size of errors in these quantities.

- When the distribution of the observations is normal with s.d. σ, the standard error of the sample median is approximately $1.25\sigma/\sqrt{n}$. (That this is 25 per cent larger than the standard error of the sample mean is one reason for preferring the latter over the sample median.)

- When the distribution of the observations is normal with s.d. σ, the standard error of the sample standard deviation (s) is $\sigma/\sqrt{2n}$.

The fact that these standard errors[35] depend upon the shape of the distribution is one of the reasons why their study is omitted at this stage.

135. **The Central Limit Theorem.** With mathematical niceties omitted, this states:

> The distribution of the sum of n independent random variables tends to the normal distribution as n tends to ∞.

Naturally, the expectation of this distribution is the sum of the expectations, and the variance is the sum of the variances.

136. Thus, if we take random samples of size n from a population whose mean is μ and whose s.d. is σ, then provided that n is reasonably large, the sampling distribution of $\bar{x}$ has the following properties:

---

[35] Notice, by the way, that we are applying the phrase "standard error" to any method of estimating anything — not just to the use of the sample mean to estimate the true mean. The precise way that "standard deviation" and "standard error" are used is often not clear to students. Roughly, "standard deviation" is used when we are thinking of a distribution, and "standard error" when we are thinking of a property of a method of estimation.

1. The expectation is μ.
2. The standard deviation is $\sigma/\sqrt{n}$.
3. The shape is normal.

Notice that the theorem said nothing about the distribution of the individual observations — this can be quite a weird, non-normal, shape, and still the distribution of the sample mean will be normal, provided that n is large enough. (Actually, there are some mathematical conditions that the distribution must obey, but they need not concern us.) This theorem is the chief reason for the great importance of the normal distribution in statistics.

137. We know (from paragraph 126) that if we add together two normally-distributed independent random variables, the result is normally distributed. We know from the Central Limit Theorem that if we add together many independent random variables, the result is normally distributed. Hence it is reasonable to conclude that if we add together a few independent random variables whose distributions are not far from normal in shape, then the result will be very close to having a normal distribution.

138. **Demonstration of the use of a table of random numbers, and of the Central Limit Theorem.** I will demonstrate the Central Limit Theorem using a distribution in which there are only two possible observations, 0 and 1; specifically, 40 per cent are 1's and 60 per cent are 0's. I will take 30 samples of size $n = 20$ observations each from this distribution, and construct a histogram of the distribution of the sample means.

139. A typical table of random numbers consists of integers between 0 and 9, each of which occurs with equal probability, as follows:

5	7	2	4	9	3	1	0	5	8	0	7	6	2	7	8	6	8	7	7
5	9	2	9	2	9	6	1	3	1	1	1	9	7	6	6	9	8	3	2
0	0	3	4	0	0	5	1	9	6	6	1	7	3	8	0	2	4	0	1
4	1	5	9	1	9	1	6	6	9	1	0	5	4	5	2	8	5	2	2
9	6	0	1	5	0	7	2	2	2	2	9	0	6	9	4	7	4	4	7
0	8	1	6	6	5	6	7	4	3	6	5	2	6	1	0	6	8	3	3
8	4	8	5	3	4	3	7	9	3	4	5	0	1	5	2	3	8	4	4
1	5	1	4	5	4	7	2	8	7	2	3	7	9	1	2	0	9	1	8

9	3	6	1	2	7	9	8	1	9	7	8	7	4	0	7	1	0	8	5
5	5	0	1	6	8	5	6	6	9	9	6	9	8	1	7	8	9	0	5
8	7	5	0	4	6	6	4	5	3	7	8	9	8	6	5	1	9	7	1
0	6	9	9	1	9	8	5	6	2	9	0	2	6	1	3	0	4	1	3
8	1	0	0	1	3	2	8	5	3	9	5	4	1	3	3	9	1	9	1
8	3	8	2	6	5	2	7	8	3	2	9	6	2	7	5	9	8	3	9
7	7	9	2	1	9	7	0	7	1	3	8	2	4	9	8	8	6	9	2
6	3	0	4	0	0	3	1	8	4	0	2	9	2	1	6	1	8	1	3
2	5	9	0	4	6	6	2	9	1	2	4	2	8	3	8	4	1	4	8
7	9	9	5	8	9	6	1	7	1	9	2	8	7	6	9	1	8	5	7
1	4	2	4	2	4	6	7	8	6	4	3	8	7	3	5	5	4	5	3
4	1	5	1	7	6	2	6	6	0	7	6	9	8	1	6	0	8	2	2
6	3	6	8	1	4	8	2	0	2	6	7	1	5	8	9	1	4	4	9
2	9	1	5	9	1	8	5	0	9	3	2	5	3	2	7	4	0	7	1
3	9	5	5	7	5	5	2	6	6	4	0	3	6	7	8	4	1	3	0
7	7	7	4	9	0	3	2	8	3	9	0	7	2	4	2	1	0	2	9
1	8	9	7	8	1	2	4	6	7	6	3	0	3	0	4	8	3	1	7
6	7	3	2	9	6	7	5	0	5	6	3	7	7	9	2	0	7	9	1
2	5	8	2	2	0	4	2	0	2	0	9	8	2	3	9	1	7	4	6
3	1	7	5	3	7	4	1	1	4	5	8	8	2	9	7	4	1	0	5
5	4	8	6	5	2	2	8	6	5	7	3	2	0	3	2	1	7	0	2
6	9	8	6	4	4	0	1	8	2	8	0	4	2	2	8	1	9	0	2

(Just as each digit 0, 1, ... 9 occurs with equal probability, so each pair of digits 00, 01, ... 99 occurs with equal probability, each triplet of digits 000, 001, ... 999 occurs with equal probability, etc.) Our first task is to change the random number table so that it has the properties we want. In this case, a convenient way is to change all occurrences of 0, 1, 2, and 3 into 1's, and all occurrences of other numbers into 0's.[36] So the observations of 0's and 1's are:[37]

0	0	1	0	0	1	1	1	0	0	1	0	0	1	0	0	0	0	0	0	0.30
0	0	1	0	1	0	0	1	1	1	1	1	0	0	0	0	0	0	1	1	0.45
1	1	1	0	1	1	0	1	0	0	0	1	0	1	0	1	1	0	1	1	0.60
0	1	0	0	1	0	1	0	0	0	1	1	0	0	0	1	0	0	1	1	0.40
0	0	1	1	0	1	0	1	1	1	1	0	1	0	0	0	0	0	0	0	0.40
1	0	1	0	0	0	0	0	0	1	0	0	1	0	1	1	0	0	1	1	0.40
0	0	0	0	1	0	1	0	0	1	0	0	1	1	0	1	1	0	0	0	0.35
1	0	1	0	0	0	0	1	0	0	1	1	0	0	1	1	1	0	1	0	0.45

[36] For other examples of the use of a table of random numbers, see Table 4.

[37] Of course, the 0's and 1's are outcomes of binomial trials for which $p = 0.40$; and the number of 1's in each row (6, 9, 12, 8, etc.) are observations of the binomial distribution with $n = 20$ and $p = 0.40$.

```
0 1 0 1 1 0 0 0 1 0 0 0 0 0 1 0 1 1 0 0    0.35
0 0 1 1 0 0 0 0 0 0 0 0 0 0 1 0 0 0 1 0    0.20
0 0 0 1 0 0 0 0 0 1 0 0 0 0 0 0 1 0 0 1    0.20
1 0 0 0 1 0 0 0 0 1 0 1 1 0 1 1 1 0 1 1    0.50
0 1 1 1 1 1 1 0 0 1 0 0 0 1 1 1 0 1 0 1    0.60
0 1 0 1 0 0 1 0 0 1 1 0 0 1 0 0 0 0 1 0    0.35
0 0 0 1 1 0 0 1 0 1 1 0 1 0 0 0 0 0 0 1    0.35
0 1 1 0 1 1 1 1 0 0 1 1 0 1 1 0 1 0 1 1    0.65
1 0 0 1 0 0 0 1 0 1 1 0 1 0 1 0 0 1 0 0    0.40
0 0 0 0 0 0 0 1 0 1 0 1 0 0 0 0 1 0 0 0    0.20
1 0 1 0 1 0 0 0 0 0 0 1 0 0 1 0 0 0 0 1    0.30
0 1 0 1 0 0 1 0 0 1 0 0 0 0 1 0 1 0 1 1    0.40
0 1 0 0 1 0 0 1 1 1 0 0 1 0 0 0 1 0 0 0    0.35
1 0 1 0 0 1 0 0 1 0 1 1 0 1 1 0 0 1 0 1    0.50
1 0 0 0 0 0 0 1 0 0 0 1 1 0 0 0 0 1 1 1    0.35
0 0 0 0 0 1 1 1 0 1 0 1 0 1 0 1 1 1 1 0    0.50
1 0 0 0 0 1 1 0 0 0 0 1 1 1 1 0 0 1 1 0    0.45
0 0 1 1 0 0 0 0 1 0 0 1 0 0 0 1 1 0 0 1    0.35
1 0 0 1 1 1 0 1 1 1 1 0 0 1 1 0 1 0 0 0    0.55
1 1 0 0 1 0 0 1 1 0 0 0 0 1 0 0 0 1 1 0    0.40
0 0 0 0 0 1 1 0 0 0 0 1 1 1 1 1 1 0 1 1    0.50
0 0 0 0 0 0 1 1 0 1 0 1 0 1 1 0 1 0 1 1    0.45
```

At the end of each row is shown the mean of the observations in that row. The frequency distribution of these means is:

0.20–0.25	0.30–0.35	0.40–0.45	0.50–0.55	0.60–0.65
3	9	10	5	3

And a rough histogram of these frequencies looks like this:

```
                  XX
         XX       XX
         XX       XX
         XX       XX
         XX       XX
         XX       XX       XX
         XX       XX       XX
XX       XX       XX       XX       XX
XX       XX       XX       XX       XX
XX       XX       XX       XX       XX

0.20     0.30     0.40     0.50     0.60
0.25     0.35     0.45     0.55     0.65
```

Despite the fact that our starting distribution of 0's and 1's was very far from normal, and despite there being only 20 observations contributing to each mean, the shape of the normal distribution is already starting to emerge.

140. The Central Limit Theorem can be used to answer questions like the following. What is the probability than the error will be less than 1.5, when the mean of a sample of size $n = 36$ is used to estimate the mean of a population whose standard deviation is 18?

> The sample mean:
>
> > Has an expectation of μ;
> >
> > Has a standard error of $18/\sqrt{36} = 3$.
> >
> > Has a normal distribution (provided we can assume that 36 is a large enough sample for the Central Limit Theorem to work its magic).
>
> The question is asking about the probability of the sample mean falling between $\mu - 1.5$ and $\mu + 1.5$. Converting these two values to z values, we find -0.5 and 0.5. Using Table 1, the required answer is $2 \times 0.1915 = 0.3830$.

141. Continuing this example, how large a sample will we need to take if we require that an error of 1.5 or more has a probability of only 10 per cent?

> Remember from an example in paragraph 105 that we first go from the probability to a z value. Using a detailed table of the normal distribution, we find that the z that corresponds to $\Phi = 0.95$ is 1.64. This must equal $\frac{1.5}{18/\sqrt{n}}$. Hence $n = 387$.

142. **What is a good method of estimation?** Various methods might be available for estimating a statistic. For example, when estimating the mean μ of a normal distribution, we might think of using the sample mean, the sample median, the average of the smallest and largest observations, and various other methods. How can we choose between them?

- Their standard error is one important guide: other things being equal, we will prefer a method that has a small standard error.

Table 4. Four further examples of the use of a table of random numbers.

Two examples of obtaining random observations from discrete distributions will be given, and then two examples of obtaining random observations from continuous distributions.

- To simulate the outcomes when rolling a dice, the convenient thing to do is to let 1, 2, 3, 4, 5, 6 represent these respective outcomes, and throw away the other numbers.

  So the first row of the random numbers in paragraph 139 would become 5 2 4 3 1 5 6 2 6, with 11 numbers having been discarded.

- Suppose we wish to simulate the following distribution:

x:	0	1	2	3	4	5	6	7
$\Pr\{X = x\}$:	0.15	0.05	0.13	0.24	0.21	0.13	0.05	0.04

  The thing to do here is to take the numbers from the table of random numbers in pairs, and to convert them to values of x using the following table:

Random numbers:	00–14	15–19	20–32	33–56	57–77	78–90	91–95	96–99
x:	0	1	2	3	4	5	6	7

  Notice how 15 possible pairs of digits have been mapped to $x = 0$, 5 have been mapped to $x = 1$, and so on.

  So the first row of the random numbers in paragraph 139 would be read as 57 24 93 10 58 07 62 78 68 77, and then these would become 4 2 6 0 4 0 4 5 4 4.

- Suppose we wish to simulate observations from an exponential distribution with mean = 1. Let the cumulative probability distribution of this be F. From paragraph 99, we know that $F = 1 - e^{-x}$. The technique here is to take observations that are uniformly distributed...

Table 4 (continued).

... over the range 0 to 1, interpret them as F, and find x from $x = -\ln(1-F)$ (where ln means the natural logarithm, i.e., the logarithm to base e). To obtain observations that are uniformly distributed over the range 0 to 1 from a table of random integers such as that in paragraph 139, read them in groups of 4, put a decimal point in front of them, and add on .00005.

So the first row of the random numbers in paragraph 139 would be read as .57245 .93105 .58075 .62785 .68775. Transforming these using $x = -\ln(1 - F)$, they become 0.8497 2.6744 0.8693 0.9885 1.1640.

- Suppose we wish to simulate observations from a normal distribution with mean = 50 and s.d. = 7. We begin as we did for the exponential distribution, reading the table of random integers in such a way that we get observations that are uniformly distributed over the range 0 to 1. Then we convert to observations from a standard normal distribution by interpreting these uniformly-distributed numbers as values of Φ, and reading off from a table of the normal distribution what values of z correspond. Finally, we convert to values of x using $x = 50 + 7z$.

Using the first row of the random numbers in paragraph 139, the three stages of our calculations are shown below.

Φ:	.57245	.93105	.58075	.62785	.68775
z:	0.1826	1.4837	0.2038	0.3262	0.4895
x:	51.28	60.39	51.43	52.28	53.43

(Most tables of the normal distribution will not give z to four decimal places; computer calculation was used to get these values.)

(Strictly speaking, the numbers obtained in the first stage of the last two examples do not have the continuous uniform distribution — it can easily be seen that numbers like .00005 and .00015, etc., can be observed, but not .00002 or .00019, for example. The error arising from this is nearly always trivial, however.)

As mentioned in paragraph 134, when sampling from a normal distribution and estimating its μ, the smaller standard error of the sample mean than of the sample median is a reason for preferring the former.

- Equally important, or more so, is that the method should be *unbiased.* This means that its expectation is the true value.
  - We can easily demonstrate that the sample mean is an unbiased method of estimating the population mean μ: $E(\bar{x}) = E\left(\frac{1}{n}\Sigma\, x\right) = \frac{1}{n}E\,(\Sigma\, x) = \frac{1}{n}\,(\Sigma\, E(x)) = \frac{1}{n}\,(\Sigma\,\mu) = \frac{1}{n}.n\mu = \mu.$
  - In paragraph 15 and subsequently, we divided by $n-1$ when calculating the sample standard deviation, rather than by n. The reason for this is that $\frac{1}{n-1}\Sigma\,(x-\bar{x})^2$ is an unbiased estimator of the population variance σ^2, whereas $\frac{1}{n}\Sigma\,(x-\bar{x})^2$ is a biased estimator.[38]

---

[38]We can prove this from what we know about the variance of $\bar{x}$. First, we write

$$\begin{aligned}
&\Sigma\,[X - E(X)]^2 \\
&= \Sigma\,[X - \bar{x} + \bar{x} - E(X)]^2 \\
&= \Sigma\,(X-\bar{x})^2 + 2\Sigma\,(X-\bar{x})[\bar{x} - E(X)] + \Sigma\,[\bar{x} - E(X)]^2 \\
&= \Sigma\,(X-\bar{x})^2 + \Sigma\,[\bar{x} - E(X)]^2 \\
&\quad \text{(because } \Sigma\,(X-\bar{x}) = 0) \\
&= \Sigma\,(X-\bar{x})^2 + n[\bar{x} - E(X)]^2.
\end{aligned}$$

Taking expectations,

$$\begin{aligned}
E(\Sigma\,[X-E(X)]^2) &= E(\Sigma\,(X-\bar{x})^2) + nE([\bar{x} - E(X)]^2) \\
nV(X) &= E(\Sigma\,(X-\bar{x})^2) + n\frac{V(X)}{n} \\
E(\Sigma\,(X-\bar{x})^2) &= (n-1)V(X),
\end{aligned}$$

and therefore

$$E\left(\frac{1}{n-1}\Sigma\,(X-\bar{x})^2\right) = V(X).$$

Notice, however, that this has referred to *variances*, not to *standard deviations.* Neither the square root of $\frac{1}{n-1}\Sigma\,(x-\bar{x})^2$ nor the square root of $\frac{1}{n}\Sigma\,(x-\bar{x})^2$ is an unbiased estimator of the standard deviation. (To try to find one is very difficult.)

- A method that is much used for devising formulae for estimating parameters[39] is the *maximum likelihood* method. This refers to finding the value of the parameter such that the likelihood of getting the data that in fact was observed is greater than for any other value of the parameter.[40]

143. **Standard error of a proportion.** Suppose we use $\hat{p} = \frac{\text{Number of successes}}{n}$ as our estimate of p, the probability of success in a binomial distribution. (As before, we use a hat ˆ to indicate an estimated quantity.) Then the standard error of $\hat{p}$ is

$$\sqrt{\frac{p(1-p)}{n}}, \tag{45}$$

[39] "Parameter" is simply a term for some quantity that we are trying to estimate — e.g., the mean μ, the s.d. σ, or the probability of success p.

[40] A convenient example is provided by the estimation of a proportion. Suppose we have an observation from a binomial distribution: perhaps we have seen 10 successes in 100 trials. How are we to estimate p, the probability of success? Intuitively, $\frac{10}{100} = 0.10$ is the way to do it. Let us now demonstrate the maximum likelihood property of this method.

- For any p, the likelihood of getting exactly 10 successes is $\binom{100}{10} p^{10}(1-p)^{90}$.

  If this is calculated assuming $p = 0.10$, the answer 0.132 is found.

  If $p = 0.09$ is assumed instead, the answer is 0.124.

  If $p = 0.11$ is assumed, the answer is 0.125.

- So it does rather look as if the likelihood of the data is highest for $p = 0.10$. We can prove this algebraically: we differentiate the likelihood with respect to p, set it equal to 0, and solve for p.

- It is actually more convenient to differentiate the logarithm of the likelihood. Call this L.

$$\begin{aligned} L &= \ln\binom{100}{10} + \ln p^{10} + \ln(1-p)^{90} \text{ (where ln means the natural logarithm, i.e., the logarithm to base } e) \\ &= \text{a constant } + 10\ln p + 90\ln(1-p) \\ \frac{dL}{dp} &= \frac{10}{p} - \frac{90}{1-p} \\ &= 0 \text{ at the maximum of } L. \end{aligned}$$

From $\frac{10}{p} = \frac{90}{1-p}$, we find $p = 0.10$.

which is usually estimated by substituting $\hat{p}$ in place of p. (Some books recommend changing n into $n-1$ when using $\hat{p}$ in place of p; Table 6 will explain this further.)

- For example, a market research company separately asks 400 consumers, who do not know each other, whether they prefer brand A or brand B of bread. When 180 prefer A, the proportion is calculated as $\frac{180}{400} = 0.45$ (or 45 per cent), with standard error $\sqrt{\frac{0.45\times0.55}{400}} = 0.025$ (or 2.5 per cent).

- If 250 students are asked whether they prefer Professor Whitehair or Dr. Greybeard as a lecturer, and 150 vote for W and 100 vote for G, we estimate the proportion preferring W to be 0.6; and we might estimate the standard error to be $\sqrt{\frac{0.6\times0.4}{250}}$, which is 0.031. However, in order for the calculation of the standard error to be valid, the different observations (that is, the different expressions of preference) need to be independent. My recollection of my student days is pretty hazy, but I do remember discussing with my friends the merits and demerits of our lecturers — our opinions certainly were not independent of each other. The standard error in such a situation is likely to be greater than that calculated by the formula — effectively, the sample size is smaller than n.

144. **Introduction to confidence intervals and testing hypotheses.** By now, we know how much a sample mean or a sample proportion is likely to vary from sample to sample. We are ready to:

- Calculate limits within which we are confident a mean, or a proportion, lies;

- Test whether a mean, or a proportion, could be some particular hypothesised value.

Some people hold strong opinions as to whether confidence intervals or hypothesis tests are the most helpful approach to statistical inference. Myself, I think they are conveying the same message in slightly different languages. (See paragraph 182 for some comments favouring confidence intervals.) Most statistics courses place slightly greater emphasis on hypothesis testing than on confidence intervals, and I will, too. The ordering of the material will as follows:[41]

---

[41] The order in which this material is taught varies quite a lot between instructors.

- Some introduce hypothesis tests before confidence intervals, rather than take C.I.'s first, as I have done.
- Some explain all the examples that use the normal distribution before those that use the t-distribution. I have chosen instead to deal with the methods for measurement data before those for counted data, and have introduced the t-distribution quite early.
- Some would prefer the "how to do it" paragraphs to be clearly separated from those offering comments on interpretation. But I think it is easier to appreciate the comments if they come in small doses.

More advanced are hypothesis tests about b, a, and r in the linear regression context, and these come at the end.

(The titles of the paragraphs are not always exactly the same as the short descriptions above.)

145. **Confidence intervals: the basic idea.** Up to now, our estimates have been single numbers; the jargon for this is "point estimates". In contrast, we might wish to give a band (or range, or interval) of values within which we are pretty confident the true value lies. This is termed "interval estimation". A confidence interval is of the form

$$\begin{matrix}\text{Best}\\\text{estimate}\end{matrix} \pm \begin{matrix}\text{Some}\\\text{number}\end{matrix} \times \begin{matrix}\text{Standard error}\\\text{of best estimate.}\end{matrix} \qquad (46)$$

(We multiply the standard error by the "some number" before adding this term to, or subtracting it from, the best estimate.) That is, it is a band, or range, of values that extends from

$$\begin{matrix}\text{Best}\\\text{estimate}\end{matrix} \textit{ minus } \begin{matrix}\text{Some}\\\text{number}\end{matrix} \times \begin{matrix}\text{Standard error}\\\text{of best estimate}\end{matrix}$$

up to

$$\begin{matrix}\text{Best}\\\text{estimate}\end{matrix} \textit{ plus } \begin{matrix}\text{Some}\\\text{number}\end{matrix} \times \begin{matrix}\text{Standard error}\\\text{of best estimate.}\end{matrix}$$

The "some number" depends on how confident we want to be. It will need to be a big number if we insist upon being highly confident, and this will lead to a wide confidence interval. It will be a smaller number if we are content with being not so confident, and this will lead to a narrower confidence interval.

146. For our first example of confidence intervals, and for many others, the "some number" will be taken from the table of the normal distribution. When making an observation of a normally-distributed random variable,

> In 90 per cent of cases, we will be within a distance of 1.64 standard deviations from the mean,

In 95 per cent of cases, we will be within a distance of 1.96 standard deviations from the mean,

In 99 per cent of cases, we will be within a distance of 2.58 standard deviations from the mean,

and so on. (We find these figures from a table of the normal distribution.)

147. **Constructing a confidence interval for one mean (large sample or known s.d.).** Here, the "best estimate" is the sample mean, and the standard error is the known s.d. divided by $\sqrt{n}$. So, if we have to calculate the 95 per cent C.I. for the mean of a population whose s.d. is known to be 2.5, and the sample mean is 69.2 for a sample of 11, the required C.I. is

$$69.2 \pm 1.96 \times \frac{2.5}{\sqrt{11}},$$

which is 69.2 ± 1.5. That is, it extends from 67.7 to 70.7. (Your instructor may want you to write this in a particular way, e.g., as 69.2 ± 1.5, as 67.7–70.7, or as $\{67.7, 70.7\}$.) Further,

- The 99 per cent C.I. is $69.2 \pm 2.58 \times \frac{2.5}{\sqrt{11}}$, which is from 67.3 to 71.1.
- The 90 per cent C.I. is $69.2 \pm 1.64 \times \frac{2.5}{\sqrt{11}}$, which is from 68.0 to 70.4.

When presenting your results, make sure your audience can easily understand what you have done — that is, that the answers to the two questions in paragraph 133x can easily be seen.

148. The s.d. may be known because there is past experience with this variable, and it is known how much variation there typically is; or, the s.d. may have been estimated from the sample, but the sample size is so large that the s.d. can effectively be assumed to be known — any errors from its perhaps being too big are cancelled out by those from its perhaps being too small. (A sample size of 30 is large enough for many purposes, and 100 for practically all.)

149. **Interpretation of a confidence interval.** Randomness means we can never be completely sure of anything. A confidence interval enables us to make a quantitative statement of our

uncertainty. If we were to take another random sample of observations, and re-calculate the C.I., we would get a slightly different sample mean, and hence a slightly different C.I. For any C.I. that we have obtained — the first one, the second one, or any other — we do not know whether the true mean is within the C.I. or not. What we can say is that if we were to repeatedly go through the whole process of taking a random sample, calculating the best estimate, and constructing the C.I., then in 95 per cent of cases (or 99 per cent, or whatever), we would have successfully captured the true mean within the C.I.[42]

150. **Calculation of the sample size required in order to meet a given specification.** Suppose we will be required to construct a 95 per cent C.I. for the mean that extends only ±0.5 unit from the sample mean, when the s.d. of the population is known to be 2.5; what sample size is required?

The specification that we have to meet is that the C.I. be

$$\text{sample mean} \pm 0.5;$$

we recall that the formula for a 95 per cent C.I. is

$$\text{sample mean} \pm 1.96 \times \text{standard error}.$$

Therefore,

$$\begin{aligned} 1.96 \times \text{standard error} &= 0.5 \\ 1.96 \times \frac{2.5}{\sqrt{n}} &= 0.5 \\ \sqrt{n} &= 9.8 \\ n &= 96. \end{aligned}$$

[42] Does this mean that "the probability that the true value is within the C.I. is 95 per cent"? To attempt a full answer to this would involve deep questions in the philosophy of probability. But I think that most statistics instructors will answer "no", and will mark your answer wrong if you say this. Stick to the explanation in paragraph 149 is my advice.

Does it mean that "we are 95 per cent confident that the true value is within the C.I."? Saying "95 per cent confident" is a very woolly, pretty meaningless, statement that will neither gain you credit nor lose it with most instructors.

151. **Hypothesis testing: the basic idea.** The idea behind a hypothesis test is as follows:

- State a particular hypothesis (e.g., about what a mean is, or what a proportion is);
- On the assumption that the hypothesis is true, determine the probability of getting a result at least as far away from the hypothesised value as the observed value is;
- If this probability is very low, reject the hypothesis.

152. **Hypothesis test for one mean (large sample or known s.d.).** Suppose that an accumulation of shells is found during excavations for building works. It might be a natural shell bed, or it might be an Aboriginal shell midden.

> Among the criteria used to judge between these alternatives is the distribution of the sizes of the shells — natural beds have all sizes, middens only have ones large enough to be used as food. (Other criteria include what shellfish species are present, the condition of the shells, and whether materials other than shell are found with the deposit.)
>
> See Attenbrow (1992) for data of this type.

It includes 9 shells of a particular species of shellfish, and their average size is 16 mm. A large collection of shellfish of this species, found in known middens nearby, averaged 30 mm. in size, with a s.d. of 12 mm. Test the hypothesis that the sample of 9 came from a population whose mean was 30 and whose s.d. was 12.

1. H_0: $\mu = 30$ (the null hypothesis)
   H_A: $\mu \neq 30$ (the alternative hypothesis)

2. If H_0 is true, the sample mean $\bar{x}$ will have a normal distribution with mean 30 and standard deviation (that is, the standard error of the mean) $12/\sqrt{9}$. As earlier in paragraph 102, it helps to draw a little sketch: in the present context, of the distribution that arises when the null hypothesis is true.

3. Consequently,

$$\frac{\bar{x} - 30}{12/\sqrt{9}}$$

will have a standard normal distribution — i.e., it can be interpreted as a z statistic. In this example, $\bar{x}$ is 16, so the result is $\frac{16-30}{12/\sqrt{9}}$, which is -3.50. (This is known as a *test statistic*, meaning a quantity whose value determines whether we reject or do not reject the null hypothesis.)

4a. From a table of the normal distribution, we find the probability of being at least as far away from the mean as -3.50 is (that is, the probability of being greater than 3.50 or less than -3.50), is .0005.

5a. Because .0005 is such a low probability, we reject the original hypothesis H_0, and prefer the alternative hypothesis, H_A. That is, the data is not consistent with the idea that $\mu = 30$. The difference from what was hypothesised is said to be "statistically significant".

Actually, it is usual to modify steps 4a and 5a as follows. We choose a "significance level" (α is the usual symbol). We decide that if the probability is less than this, we will reject H_0. Now, instead of working out the probability (.0005) corresponding to the observed z, we work out what value of z would correspond to the chosen value of α. Steps 4a and 5a are therefore replaced by the following:

4b. We choose a significance level of 0.05 (that is, 5 per cent). Corresponding to a total probability of 0.05 being in the tails of the distribution (that is, in the extreme ends) is a z of 1.96.

5b. Because the observed z lies outside the range -1.96 to 1.96, we reject the original hypothesis H_0, and prefer the alternative hypothesis, H_A. That is, the data is not consistent with the idea that $\mu = 30$.

This is easier because there are only a few values of the significance level α that are commonly used, and one soon gets to remember the corresponding values of z. (They are known as *critical values*.) If the observed z had turned out to be 1.92, for example, we would not have rejected H_0. (Some instructors, by the way, object to the wording "accept" H_0 in such a situation.)

153. **Choice of significance level.** The choice of α is, it has to be admitted, somewhat arbitrary. The most common choice is $\alpha = 0.05$, though $\alpha = 0.01$, $\alpha = 0.10$, and others, are sometimes used. But it is no more arbitrary than the choice of degree of confidence when constructing a C.I.

- If we choose a very small α (say, 0.001), it is quite likely we will fail to reject H_0 even if μ is appreciably different from the hypothesised value. If we choose a large α (say, 0.20), a rejection of H_0 could well occur when μ does equal the hypothesised value, or at any rate is so close to it that the difference is of no practical importance.

- Notice, by the way, the distinction between *statistical* and *practical* significance. A statistically-significant difference may be of no practical significance, especially if the difference is small in magnitude, and the statistical significance arose because of a large sample size; a difference that fails to be of statistical significance may in fact reflect a true difference that is sufficiently large to be important, especially if the sample size was small and the standard error used in the hypothesis test therefore quite large.

154. **The crucial calculation.** A similar formula to that above will appear in many more hypothesis tests. The crucial calculation is that of the difference between observed and hypothesised values, expressed in units of the standard error; that is,

$$\frac{\text{Observed} - \text{Hypothesised}}{\text{Standard error}}. \tag{47}$$

155. **Two-sided and one-sided alternative hypotheses.** In the above example, we would have rejected the null hypothesis H_0 had the observed result been sufficiently far from the hypothesised result *in either direction.* We expressed this as the alternative hypothesis H_A: $\mu \neq 120$. Sometimes, we may wish to make it easier

to reject the null hypothesis when the data deviates from the null hypothesis in the expected direction — at the expense of making it impossible to reject the null hypothesis if the data deviates from the null hypothesis in the opposite direction to that expected. For example, we may be sure that only one direction of deviation from the null hypothesis is possible.

156. Suppose we knew that the mean size of the 9 shells in the previous example would not exceed 30 mm. For instance, we could say there is a possible reason for the mean being less than 30 mm. (namely, that the shells came from a natural bed, not a midden), but there is no possible reason why it should be greater than 30 mm. Then the hypothesis test would proceed as follows.

1. H_0: $\mu = 30$ (the null hypothesis)[43]
   H_A: $\mu < 30$ (the alternative hypothesis)

2. If H_0 is true, the sample mean $\bar{x}$ will have a normal distribution with mean 30 and standard deviation (that is, the standard error of the mean) $12/\sqrt{9}$.

3. Consequently,

   $$\frac{\bar{x} - 30}{12/\sqrt{9}}$$

   will have a standard normal distribution — i.e., it can be interpreted as a z statistic. In this example, $\bar{x}$ is 16, so the result is $\frac{16-30}{12/\sqrt{9}}$, which is -3.50.

4. We choose a significance level of 0.05 (that is, 5 per cent). Corresponding to a probability of 0.05 being in the left-hand tail of the distribution, the critical value of z is -1.64.

5. Because the observed z is less than -1.64, we reject the original hypothesis H_0, and prefer the alternative hypothesis, H_A. That is, the data is not consistent with the idea that $\mu = 30$.

157. I find it helpful to illustrate the regions within which one rejects or does not reject the null hypothesis as follows.

---

[43] In the present context of a one-sided alternative hypothesis, some instructors will want this written as H_0: $\mu \geq 30$.

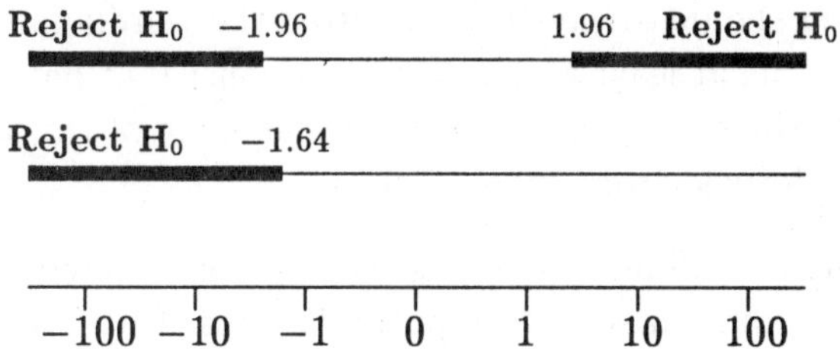

In the upper part of the picture is shown the region where one rejects H_0 when the alternative hypothesis is two-sided, H_A: $\mu \neq 30$. Below it is shown the region where one rejects H_0 when the alternative is the one-sided hypothesis H_A: $\mu < 30$.

158. Some further notes about one-sided alternatives:

- The decision on a particular alternative hypothesis must be made *before* knowing what the data is. To do otherwise is plain cheating.

  Suppose the mean shell size had worked out to be 23, instead of 16, and therefore the observed z was -1.75. This is not statistically significant at the 0.05 level of significance, when performing a two-sided test. But it is when performing a one-sided test. Very tempting if one wants to find a difference!

- There is a serious line of argument that says one should virtually never use a one-sided H_A. Proponents of this line will say that it is virtually never that one direction of deviation from the null hypothesis is impossible. And they will add that the following seems ridiculous:

  Suppose the one-sided H_A: $\mu < 30$ is specified.

  But further suppose the very surprising result $\bar{x} = 40$ (which leads to $z = \frac{40-30}{12/\sqrt{9}} = 2.50$) is observed.

  Despite the very surprising result, one must not reject H_0 — the data is even more inconsistent with H_A than it is with H_0.

- In an examination, whether a two-sided or a one-sided test is required should always be clear from the wording of a question. If in doubt, at least remember to be consistent between the H_A chosen, the α chosen, and the critical value of the test statistic that is derived from these choices.

- One-sided confidence intervals can be constructed, analogously to performing one-sided hypothesis tests. But I will not spend space writing out the details.

159. **Type I and Type II errors.** In reality, the null hypothesis H_0 is either true or false. Our conclusion at the end of a hypothesis test is to reject or not to reject the null hypothesis. There are thus two types of error, and two types of correct decision, that can be made in hypothesis tests.

	Our decision	
	Do not reject H_0	Reject H_0
The true state of nature:		
H_0 is correct	*Correct*	**Type I error**
H_0 is false	**Type II error**	*Correct*

- If the conclusion of our hypothesis test is that we reject H_0 when in fact it is correct, we have made what is called a Type I error. (The probability of doing this, given that H_0 is indeed correct, is the significance level, α.)

- If the conclusion of our hypothesis test is that we do not reject H_0 when in fact it is false, we have made a Type II error. (The probability of doing this, given that H_0 is indeed false, is usually given the symbol β. Naturally, it depends upon how far away from the true state of affairs is H_0.)

The penalty we pay for choosing a very low α, and thus having a very low probability of a Type I error, is that we have a high probability of a Type II error (unless we increase the sample size, n).

How to construct a statistical test that has as low a β as possible, for a given value of α, is an important subject studied within mathematical statistics, but is too advanced for us now.

160. An important criticism of the hypothesis testing tradition in statistics is that it fails to properly integrate

> The evidence from the observations (and perhaps from previous experiments)

with

> The costs and benefits of the two types of error and the two types of correct decision,

in order to

> Arrive at the best decision on future action.

Controversies over the introduction of new drugs, when the evidence is incomplete as regards both their effectiveness and their safety, exemplify this point. A technique of increasing importance is "meta-analysis". By this is meant the quantitative combination of evidence from many quite separate experiments, in order to arrive at a conclusion that is, in effect, based on a larger sample size than any of the individual experiments.

> For example, if the median survival time of cancer patients was found to be greater for those given drug D than for those not given the drug, in each of six independent studies, one would be inclined to believe the drug was beneficial, even if the effect was not statistically significant in any of the individual studies. (The probability of the difference being in the same direction in all six studies is $2.\left(\frac{1}{2}\right)^6 = \frac{1}{32}$, under the null hypothesis of no difference.)

There are both practical problems (relating to, e.g., whether the different experiments are truly comparable, what is the quality of each, and whether experiments giving a positive result are more likely to be published than those giving a negative result), and technical statistical problems here.

161. **Concepts recently met.** Here is a list of specialist statistical jargon that we have recently met:

Confidence interval
Null hypothesis
Alternative hypothesis (two-sided; one-sided)
Test statistic
Significance level
Critical value
Type I error
Type II error.

The student should be sufficiently familiar with these as to be able to explain them clearly to an intelligent person who has not had training in statistics.

162. **Constructing a confidence interval for one mean (small sample and estimated s.d.).** In paragraph 147, we were able to use the normal distribution to give us the "some number". When we estimate the s.d. from our sample, we cannot do this.[44] We take the "some number" from the t-distribution, instead.[45] See Table 5. Actually, there are lots of different t-distributions; the one you choose is the one appropriate to your sample size. The distributions appropriate to large sample sizes are very similar to the normal distribution; those for smaller and smaller sample sizes get further and further away from the shape of the normal distribution. Tables of the t-distribution are labelled by the number of *degrees of freedom.* This is usually given the symbol ν (the Greek letter nu). In the present context, the number of degrees of freedom is one less than the number in the sample; that is, $\nu = n - 1$. So, if we have to calculate the 95 per cent C.I. for the mean, when the sample mean

[44] The basic reason is that if we have a normally-distributed quantity (the "best estimate"), and we add on to it a constant (the "some number" times a known standard error), the result will still be normally-distributed. But if the s.d. is not known, being merely estimated from our sample, the s.d. constitutes an additional source of variability: we are computing (Best estimate) $\pm$ (Some number) $\times$ (A random variable). To achieve the required level of significance, we have to change the "some number" that we are multiplying by.

[45] It is sometimes referred to as Student's t-distribution. The story behind this is that the person who derived it, W.S. Gosset (1876–1937), was employed by a commercial firm (the Guinness brewery), who insisted that he use a pen-name when publishing his work; and he chose the pen-name "Student". (He did not actually use the symbol t, that convention came later.)

Table 5. A (very short) table of the t-distribution.

ν	In this column is the value of t such that the probability of exceeding it is 0.025; that is, the probability of being less than $-t$ or greater than t is 0.05	In this column is the value of t such that the probability of exceeding it is 0.005; that is, the probability of being less than $-t$ or greater than t is 0.01
1	12.71	63.66
2	4.30	9.92
3	3.18	5.84
4	2.78	4.60
5	2.57	4.03
6	2.45	3.71
7	2.36	3.50
8	2.31	3.36
9	2.26	3.25
10	2.23	3.17
11	2.20	3.11
12	2.18	3.05
13	2.16	3.01
14	2.14	2.98
15	2.13	2.95
30	2.04	2.75
∞	1.96	2.58

(The final row — infinite degrees of freedom — refers to the normal distribution.)

and s.d. are respectively 69.2 and 2.5 for a sample of 11, the C.I. is

$$69.2 \pm 2.23 \times \frac{2.5}{\sqrt{11}},$$

which is 69.2 ± 1.7. That is, it extends from 67.5 to 70.9.

163. In summary: when the s.d. has been estimated from the sample, rather than being known in advance, we need

$$\pm\, t.\frac{s}{\sqrt{n}},$$

instead of

$$\pm\, z.\frac{\sigma}{\sqrt{n}}.$$

164. In paragraph 1, we had a dataset of 15 percentages of tin. We found (paragraph 4) that the mean was 3.15, and (paragraph 15) that the s.d. was 1.53. We can therefore find that the 95 per cent confidence interval for the mean is

$$3.15 \pm 2.14 \times \frac{1.53}{\sqrt{15}},$$

i.e., it extends from 2.30 to 4.00. Let us now show this on the box-and-whisker plot of the data (from paragraph 20).

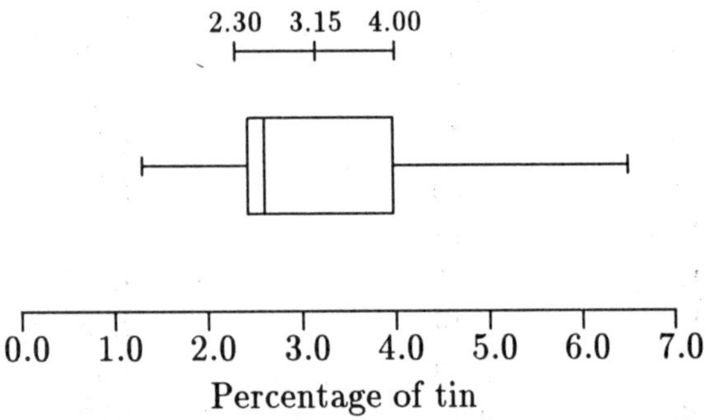

As the box accounts for the middle 50 per cent of observations, we can see that the C.I. in this case contains about half of the observations within its span. There is nothing surprising about this — the C.I. refers to the range that we are confident the population *mean* lies within, not any *individual* observation.

> The C.I. reflects the total information in the whole sample. For a large sample size, the average and the s.d. will be about the same as in a small sample; but the standard error of the mean and the C.I. for the mean will be smaller than in a small sample, reflecting the increased amount of information in the larger sample.

165. **Hypothesis test for one mean (small sample and estimated s.d.).** When constructing a confidence interval, we changed from

> Taking the "some number" from the normal distribution, when the s.d. was known,

to

> Taking it from the t-distribution, when the s.d. was estimated.

Similarly, when performing a hypothesis test in a situation where the s.d. is merely estimated, we take the critical value from the t-distribution, instead of from the normal distribution.

166. Suppose that 9 shells are measured, and it is found that their average size is 16 mm., and the s.d. is 12; test the hypothesis that the average size is 30 mm. (This is only slightly different from the problem in paragraph 152, and what we do is only slightly different.)

1. H_0: $\mu = 30$ (the null hypothesis)
   H_A: $\mu \neq 30$ (the alternative hypothesis)

2. If H_0 is true, the sample mean $\bar{x}$ will have a normal distribution with mean 30 and standard deviation (that is, the standard error of the mean) estimated to be $12/\sqrt{9}$.

3. However,
$$\frac{\bar{x} - \mu}{s/\sqrt{n}} \qquad (48)$$
   will not have a normal distribution — the denominator is a random variable, not a constant. Instead, it must be interpreted as a t statistic. (As in the analogous confidence interval setting, the number of degrees of freedom is $n - 1$.) In this example, the result is $\frac{16-30}{12/\sqrt{9}}$, which is -3.50.

4. We choose a significance level of 0.05 (that is, 5 per cent). Corresponding to a total probability of 0.05 being in the tails of the distribution is a critical value of t (with 8 degrees of freedom) of 2.31.

5. Because the observed t lies outside the range -2.31 to 2.31, we reject the original hypothesis H_0. That is, the data is not consistent with the idea that $\mu = 30$.

Comparing this result with that of paragraph 152, we see that in this case the additional uncertainty (in the standard deviation) has not changed our "reject H_0" decision. But if we had happened to find that the sample mean was 21, so that the value of our test statistic was -2.25, we would have rejected H_0 in the case of the s.d. being known to be 12, but we would not have rejected H_0 in the case of the s.d. being unknown, but estimated from the sample to be 12.

167. **Hypothesis test about a difference between paired observations (the paired t-test).** This is a variation on the previous test. It is for a situation which is apparently quite different — we have a number of *pairs* of observations, and we want to test a hypothesis about the difference between them (often, the hypothesis is that the difference is 0). But there is a simple trick to bring it into the format of the previous test. Simply calculate the differences that the pairs exhibit, and proceed as previously. This test is of wide application in the experimental sciences — for instance, the data might refer to several people, being obtained from each one both before and after the administration of some drug.

168. Suppose that the number of months elapsing between the birth of a baby and the birth of the same woman's next baby can be found; and that the average intervals between births (in months) turned out to be as below, for 6 19th-century German villages.

Village	Mean, when baby died	Mean, when baby survived
1	18	25
2	17	25
3	17	20
4	19	24
5	19	30
6	18	28

The comparison being made is between cases when the earlier baby died within the first month after birth and those when the baby survived at least one year.

> This data is indirect evidence regarding the prevalence of breastfeeding: breastfeeding tends to postpone the resumption of ovulation in women following birth, and hence lengthen the interval until the next birth; intervals following the birth of a child that dies very early in infancy would be largely unaffected, while intervals following births of infants who survive a year would reflect the full impact. If breastfeeding is rare, there will be little or no difference. (See Knodel, 1983, for study of data of this type.)

To test whether the average difference between the two conditions is 0, we proceed as follows.

1. H_0: $\mu = 0$ (where μ is the mean difference)
   H_A: $\mu \neq 0$

2. Calculate the differences for the above pairs.

   $$\begin{array}{r} -7 \\ -8 \\ -3 \\ -5 \\ -11 \\ -10 \end{array}$$

   These have mean -7.33 and s.d. 3.01.

3. If H_0 is true, the sample mean difference will have a normal distribution with mean 0 and standard deviation (that is, the standard error of the mean) estimated to be $3.01/\sqrt{6}$.

4. Calculate
   $$\frac{-7.33 - 0}{3.01/\sqrt{6}},$$
   which is 5.97. Interpret this as a t statistic with 5 degrees of freedom.

5. We choose a significance level of 0.05 (that is, 5 per cent). Corresponding to a total probability of 0.05 being in the tails of the distribution is a critical value of t (with 5 degrees of freedom) of 2.57.

6. Because the observed t lies outside the range -2.57 to 2.57, we reject the original hypothesis H_0. That is, the data is not consistent with the idea that the mean difference is 0.

Remarks:

- Suppose that it were known, from a large sample of other villages, that the usual difference in the interval between births was 12 months. To test whether the average difference for the above 6 villages is consistent with this figure, we would calculate $\frac{-7.33-(-12)}{3.01/\sqrt{6}}$. Finding the result to be 3.80, we reject the null hypothesis that the difference $= 12$.
- The construction of a C.I. is similar to that in paragraph 162, once the differences have been obtained.
- If one had the data for each individual infant, or at least had the standard errors associated with the means that have been given, one could perform a statistical test on the data for each village separately. (The method would be that of paragraphs 169–171 below.) But one might not have that data, or there might be something distinctive about these particular villages, so that their results ought to be considered together, and compared with other villages in the light of the typical size of village-to-village variation in the difference between the paired observations.

169. **Hypothesis test about a difference between the means of two samples (large samples).** Now, we turn to the comparison of the means of two groups of observations, instead of the comparison of the mean of one group with a specified value. We will test whether the difference could be a specified value; this is often 0 — i.e., it is hypothesised that there is no difference between the means of the two groups. We will again calculate

$$\frac{\text{Observed} - \text{Hypothesised}}{\text{Standard error}},$$

but now it will be

$$\frac{\text{Observed } \textit{difference} - \text{Hypothesised } \textit{difference}}{\text{Standard error } \textit{of the difference}}.$$

We will be concerned, for the present, only with the case where the sample sizes of the two groups are large enough for us to use the normal distribution, rather than the t-distribution.

170. If the standard deviations in the two groups are σ_1 and σ_2, and the sample sizes are n_1 and n_2, the standard errors of the means will be $\sigma_1/\sqrt{n_1}$ and $\sigma_2/\sqrt{n_2}$. Consequently, the standard error of the difference will be

$$\sqrt{\frac{\sigma_1^2}{n_1} + \frac{\sigma_2^2}{n_2}} \tag{49}$$

(because the variance of a difference is the sum of the variances, provided the random variables are independent — see paragraph 121). If n_1 and n_2 are large, we will be able to substitute the estimated s.d.'s s_1 and s_2 for the population values σ_1 and σ_2, with negligible loss of accuracy. Furthermore, the means will approximately have normal distributions, and hence so will their difference (paragraph 126).

171. It is possible to use measurements of cores extracted from mountain ice caps to deduce annual rainfall many hundreds of years ago: large snow layer thickness indicates heavy rainfall at lower altitudes, and small layer thickness implies lower rainfall. (See paragraph 171x for further background to this example.) Suppose we have 225 measurements of ice cap thickness from before AD 1000, and 376 measurements from after that date, and that the summary statistics are as below.

Pre-AD 1000	mean = 1.49,	s.d. = 0.39
Post-AD 1000	mean = 1.23,	s.d. = 0.32

Is the apparent difference in means a statistically-significant one?

1. H_0: $\mu_1 - \mu_2 = 0$ (that is, $\mu_1 = \mu_2$)
   H_A: $\mu_1 - \mu_2 \neq 0$ (that is, $\mu_1 \neq \mu_2$)

2. If H_0 is true, the difference between the sample means will have a normal distribution (since the sample sizes are large) whose mean is 0 and whose standard deviation (that is, the standard error of the difference) is estimated to be $\sqrt{\frac{0.39^2}{225} + \frac{0.32^2}{376}}$.

3. So,

$$\frac{(\bar{x}_1 - \bar{x}_2) - 0}{\sqrt{\frac{s_1^2}{n_1} + \frac{s_2^2}{n_2}}} \qquad (50)$$

can be taken to have a standard normal distribution when the null hypothesis is true. In this example, the observed value of the difference is 1.49 − 1.23, so the result is $\frac{1.49 - 1.23}{\sqrt{\frac{0.39^2}{225} + \frac{0.32^2}{376}}} = \frac{0.26}{0.031}$, which is 8.4. So at any reasonable level of significance, we will conclude the two population means are different. If we insist on spelling this out fully, we will write:

4. We choose a significance level of 0.05 (that is, 5 per cent). Corresponding to a total probability of 0.05 being in the tails of the distribution is a critical value of z of 1.96.

5. Because the observed z lies outside the range −1.96 to 1.96, we reject the original hypothesis H_0. That is, the data is not consistent with the idea that there has been no change in the mean ice cap thickness.

Remarks:

- There is nothing special about an expected difference of 0. If for some reason we had wished to test the null hypothesis that the difference in means was 0.20, the value of the test statistic would have been $\frac{0.26-0.20}{0.031}$, which is 1.94. This lies within the range −1.96 to 1.96, so we would not reject H_0. That is, the data is consistent with the idea that the change in mean ice cap thickness is 0.20.

- The construction of a C.I. for the difference between the means would be based on the same principles as the above test, but I will not spell out the details.

171x Background to, and comments on, the example in paragraph 171:

- The example was suggested by the paper by Ortloff and Kolata (1993). From their summary: "After approximately 700 years of growth and colonial expansion, the Tiwanaku state disintegrated as a regional political force in the south-central Andes between *c.* AD 1000–1100. This paper examines the collapse of the state of Tiwanaku through the lens of its agricultural history... We present evidence that the collapse of Tiwanaku intensive agriculture was triggered by regional change in climatic conditions recorded in highly resolved palaeoenvironmental data derived from the Quelccaya ice cap of southern Peru and in sediment cores from Lake Titicaca... [Post-AD 1000, there was] a statistically significant decrease in mean precipitation level that persisted until *c.* AD 1400." In brief, lack of rain caused the downfall of Tiwanaku.

- After performing a statistical test like that in paragraph 171, Ortloff and Kolata say: "These results indicate that the probability that the ice cap thickness distribution before AD 1000 and that after AD 1000 are from the same population is < 0.00001." This is complete rubbish, and if you write something like this in an exam, your examiner is likely to give up in despair. No doubt it was a momentary slip, and Ortloff and Kolata know perfectly well that they should have said "If the two ice cap thickness distributions came from the same population, then the probability of getting a difference at least as large as that observed is < 0.00001."

  In other words, what we are getting from a significance test is an indication of

$$\Pr\{\text{Data} \mid \text{Hypothesis is true}\},$$

not

$$\Pr\{\text{Hypothesis is true} \mid \text{Data}\}.$$

- Here is a suitable place to emphasise the necessity for an external criterion to divide a set of measurements into two groups that will be compared. Let us ask ourselves why AD 1000 was chosen as the date around which to construct a before versus after comparison. Common sense tells us that when we have a series of measurements, there are bound to be ups and downs;

and if the point where we divide them is chosen because the measurements after it seem to differ from those before it, then an ordinary statistical test is not valid. The division of the series needs to be on the basis of *external* evidence, i.e., evidence other than that provided by the measurements themselves — in this instance, perhaps documentary or archaeological evidence of economic or political decline dating from about AD 1000.

172. **Hypothesis test about a difference between the means of two samples (small samples): the Mann-Whitney U-test.** There does exist a t-test for comparing the means of two groups. However, it is quite complicated to carry out. The Mann-Whitney U-test[46] is much simpler. Furthermore, it serves as an introduction to statistical methods that rely upon the *ranks* of the observations; these throw away information about the exact values of the observations, but what they gain by this is greater robustness to the presence of outliers. (In paragraph 52, we met Spearman's rank correlation coefficient, which has this advantage as compared with the product-moment correlation coefficient.)

173. The test proceeds by ranking the observations, in both groups together, from smallest to highest. Let the numbers in the two groups be n_1 and n_2, and $n = n_1 + n_2$. The average rank is $(n+1)/2$, so we would expect the sum of the ranks of observations in the first group to be about $n_1(n+1)/2$. The formula for the standard deviation of the sum of ranks has been derived, and is $\sqrt{n_1 n_2 (n+1)/12}$. So the test statistic is

$$\frac{\text{Observed sum of ranks in group 1} - n_1(n+1)/2}{\sqrt{n_1 n_2 (n+1)/12}}, \qquad (51)$$

and is interpreted as having a standard normal distribution when the null hypothesis of no difference in the means is true.

174. Suppose observations in groups 1 and 2 were as follows:

Group 1: 3.2, 4.5, 1.7, 5.8, 3.1, 4.4, 4.0, 1.2
Group 2: 4.8, 2.3, 5.6, 6.2, 5.7, 4.3, 6.3, 5.5, 4.9, 5.2, 3.3.

[46] This is also known as Wilcoxon's sum of ranks test. But I will avoid this name because the name of Wilcoxon is associated with other tests, also.

(Here, $n_1 = 8$ and $n_2 = 11$.) When the observations are replaced by their ranks, the result is:

Group 1: 5, 10, 2, 17, 4, 9, 7, 1
Group 2: 11, 3, 15, 18, 16, 8, 19, 14, 12, 13, 6.

The sum of ranks in group 1 is 55, and its expectation under the null hypothesis of equal means is $8 \times 20/2 = 80$, with standard deviation $\sqrt{8 \times 11 \times 20/12} = 12.11$. The value of the test statistic is $\frac{55-80}{12.11} = -2.06$, which is statistically significant, when performing a two-sided hypothesis test at the 0.05 level of significance. That is, we reject the hypothesis of no difference between the means, and conclude there genuinely is a difference.

175. Remarks:

- If two or more observations are exactly the same, give them all their average rank.
- To treat the test statistic as being normally distributed is only reasonably accurate when n_1 and n_2 are both sufficiently large (say, both are 8 or bigger). If this is not so, then special tables of the critical values have to be consulted. (These can be found in many statistics textbooks.)

176. **The confidence interval for a proportion.** Here, the "best estimate" is the sample proportion $\hat{p}$, and the standard error is $\sqrt{\dfrac{\hat{p}(1-\hat{p})}{n}}$.

If 850 cars entering a city centre are observed, and 360 are found to have two or more occupants, the proportion is estimated to be $\frac{360}{850} = 0.424$, with standard error estimated to be $\sqrt{\frac{0.424 \times 0.576}{850}} = 0.017$, and the 98 per cent confidence interval extends from $0.424 - (0.017 \times 2.33) = 0.384$ up to $0.424 + (0.017 \times 2.33) = 0.464$. (There is no special reason for choosing the 98 per cent confidence level here.)

177. In the context of proportions, one always uses the normal distribution, not the t-distribution. The easiest thing to do is just to

Table 6. The connexion between the standard errors of means and of proportions.

Let us say that a car with two or more occupants is a "measurement" of 1, and a car with one occupant is a "measurement" of 0.

- Then the average $\bar{x}$ of these "measured" observations is exactly $\hat{p}$: $\frac{360\times1+490\times0}{360+490} = \frac{360}{850} = 0.424$.

- Using $s = \sqrt{\frac{\Sigma x^2 - \frac{1}{n}(\Sigma x)^2}{n-1}}$, the estimate of the standard deviation is

$$\sqrt{\frac{360 - \frac{1}{850}(360)^2}{849}} = \sqrt{\frac{360\left(1-\frac{360}{850}\right)}{849}} = \sqrt{\frac{360}{849}\times\frac{490}{850}}.$$

- And the standard error of the mean is $\frac{1}{\sqrt{n}}$ times this, i.e., $\sqrt{\frac{\frac{360}{849}\times\frac{490}{850}}{850}}$, which is exactly the estimated standard error of the proportion, apart from $\frac{360}{849}$ having replaced $\frac{360}{850}$.

  This may alternatively be written as $\sqrt{\frac{\frac{360}{850}\times\frac{490}{850}}{849}}$; because of this, some textbooks use $\sqrt{\frac{\hat{p}(1-\hat{p})}{n-1}}$ as the formula for the standard error of an estimated proportion, instead of $\sqrt{\frac{\hat{p}(1-\hat{p})}{n}}$.

- Then we would construct the C.I. using this standard error and t with 849 degrees of freedom. Because 849 is so close to 850, and t with 849 d.f. is so close to the normal distribution, our answer would be very close to that already given in paragraph 176.

accept this — proportions refer to counted data, and means refer to measured data, and these are very different things. But some of my students suggest using a t-distribution here, since what we have got is an estimated standard error. See Table 6 concerning this.

178. **Being pessimistic when planning.** Suppose we are planning a survey, the outcome of which is to be knowledge concerning a proportion. If the specification for the outcome is that the 95 per cent C.I. must be only 0.04 wide (that is, ± 0.02), what sample size is required? Using the method of paragraph 150,

$$\begin{aligned} 0.02 &= 1.96 \times \text{standard error} \\ &= 1.96\sqrt{p(1-p)/n}. \end{aligned}$$

If we know p, we can solve this equation to get n. The trouble is, we do not know p, that is what we are trying to find out!

- One thing we can do is to use our knowledge of the situation to make a guess at p.

- Alternatively, if we really have no idea what p might be, we should realise that the sample size required will be larger if $p = \frac{1}{2}$ than if p is anything else.

  Rearranging the above equation, we find $n = 98^2 p(1-p)$. Plotting n as a function of p, we obtain the graph below.

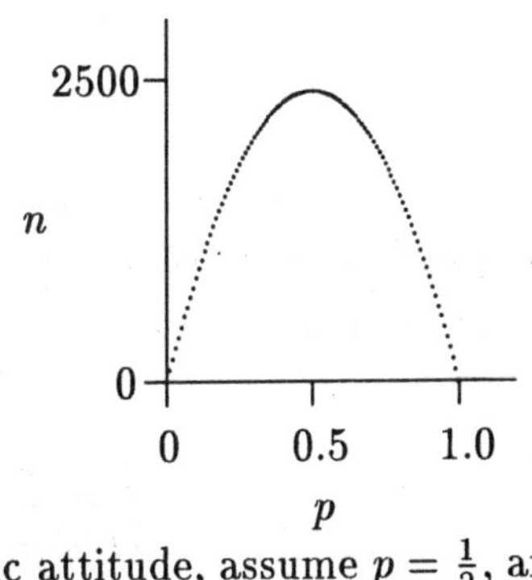

So, we take a pessimistic attitude, assume $p = \frac{1}{2}$, and work out n. Then, whatever $\hat{p}$ turns out to be, $1.96\sqrt{\hat{p}(1-\hat{p})/n}$ will be as small as we specified, or smaller.

In our example, we would find

$$
\begin{aligned}
\sqrt{n} &= \frac{1.96\sqrt{\frac{1}{2}\cdot\frac{1}{2}}}{0.02} \\
&= 49 \\
n &= 2401.
\end{aligned}
$$

179. **Hypothesis test about a proportion.** We will again calculate

$$\frac{\text{Observed} - \text{Hypothesised}}{\text{Standard error}},$$

but now it will be

$$\frac{\text{Observed } \textit{proportion} - \text{Hypothesised } \textit{proportion}}{\text{Standard error } \textit{of the proportion}}.$$

Suppose that in recent years, the proportion of young women charged with infanticide who have been convicted has been 0.40. The penalty for this crime is supposed to be execution by guillotine. Then the law is changed, giving the jury the right to convict "with extenuating circumstances", meaning that the death penalty cannot be applied. After the change of law, there are 170 trials of women accused of this crime, and 95 result in conviction. Test the hypothesis that the probability of conviction is still actually 0.40. (The paper by Donovan, 1991, includes study of statistics of this type.)

1. H_0: $p = 0.40$
   H_A: $p \neq 0.40$

2. If H_0 is true, the observed proportion will have a normal distribution (provided the sample size is large) whose mean is 0.40 and whose standard error is $\sqrt{\frac{0.40\times0.60}{170}}$.

3. Calculate

$$\frac{\frac{95}{170} - 0.40}{\sqrt{\frac{0.40\times0.60}{170}}},$$

   which is 4.23. Interpret this as having a standard normal distribution when the null hypothesis is true.

4. The observed value exceeds the critical value of z corresponding to any of the usual significance levels (e.g., 2.58 for the 0.01 level of significance).

5. And so we reject the original hypothesis H_0. That is, the data is not consistent with the idea that the proportion is 0.40.

180. Suppose we had performed the above test in terms of *numbers*, rather than *proportions*. We would have calculated $\frac{95-68}{\sqrt{170\times0.40\times0.60}}$, giving 4.23, precisely the same answer. But we know (from paragraph 108) that we should really make a continuity correction in this situation, and calculate $\frac{94.5-68}{\sqrt{170\times0.40\times0.60}}$, giving 4.15. This is the better answer; to get it when working in terms of proportions, one would calculate $\frac{\hat{p}-\frac{1}{2n}-p}{\sqrt{p(1-p)/n}}$. However, it is usual not to bother with a continuity correction when doing a hypothesis test for a proportion, because one usually has such a large sample size that it makes negligible difference.

181. **Hypothesis test for whether there is a difference between two proportions.** The only null hypothesis that we will consider in the context of comparing two proportions is that the two proportions are equal — that is, their difference is 0. Records from London for the late 17th century concerning emigrant servants enable various characteristics of these people to be studied (Grubb, 1992). Let us suppose we have data as follows:

	Number	Proportion whose father was dead
A. Male servants aged under 21	52	0.365
B. Male servants aged 21 and over	43	0.163

Decide whether the observed difference in the proportions is a statistically significant one, using the 0.01 level of significance. (On the grounds that the father's death is likely to have a greater impact on a son under 21 years old than on an older one, and therefore the likelihood of choosing the option of emigration as a servant will, if anything, be higher for younger men, specify a one-sided alternative hypothesis.)

1. H_0: $p_A - p_B = 0$
   H_A: $p_A - p_B > 0$

2. If H_0 is true, the observed difference in proportions will have a normal distribution (provided the sample size is large) whose mean is 0, and whose standard error is $\sqrt{\frac{p(1-p)}{n_A} + \frac{p(1-p)}{n_B}}$, where p is the true proportion of emigrant male servants whose father was dead, whether they were aged 21+ or whether they were younger. (Remember that the two proportions are equal if H_0 is true.)

3. Our best estimate of p is not simply the average, $(p_A + p_B)/2$, but should take into account the two sample sizes. Our best estimate of p is

$$\frac{\text{total number of those whose father was dead in the samples}}{\text{total number in the samples}},$$

   which can also be written as $\frac{n_A p_A + n_B p_B}{n_A + n_B}$. In the present case, it is $\frac{19+7}{52+43}$, which is 0.274.

4. Now we are ready to calculate

$$\frac{\text{Observed difference} - \text{Hypothesised difference}}{\text{Standard error of the difference}},$$

   and interpret it as having a standard normal distribution when the null hypothesis is true. It works out to be

$$\frac{(0.365 - 0.163) - 0}{\sqrt{\frac{0.274 \times 0.726}{52} + \frac{0.274 \times 0.726}{43}}},$$

   which is $\frac{0.202}{0.0919} = 2.20$.

5. Performing a one-sided test at the 0.01 level of significance, the critical value is 2.33.

6. And so we do not reject the original hypothesis H_0. That is, the data is consistent with the idea that the difference in proportions is 0.

182. **Comment on the decision from hypothesis tests.** Of course, the failure to reject the null hypothesis in this last example has come about because of the significance level chosen (0.01). Had $\alpha = 0.05$ been used instead, the critical value (one-sided) would have been 1.64, and we would have rejected H_0 comfortably.

- Some statisticians say that the only thing to come out of a hypothesis test is the decision as to whether to reject H_0 or not, at a previously-chosen significance level. Consequently, they will not care that we have come so close to rejecting H_0 here. Reject it we have not, and that is the end of the story.

- I think a majority of statisticians, however, will take a more relaxed view. They will be conscious that a hypothesis test on one dataset is only a part of the total picture: no doubt there are other datasets that bear upon the same question, there may be prior beliefs about the likelihood of the truth of the null hypothesis, and in any case the costs and benefits of the different wrong and right decisions should be borne in mind. Consequently, they will view the level of significance attained as being just a rough indication of the strength of evidence against H_0, and will say the α attained here is between 0.05 and 0.01.

- Relevant here are some comments by Cowgill (1975) that deprecate too close adherence to the reject or not reject philosophy, and mildly favour confidence intervals rather than hypothesis tests: "Statistical inference is never complete if we only ask whether sample results are consistent with one specific... hypothesis about the population. We always have to think also about with what other hypotheses the results may also be consistent... The upshot is that, especially for small samples, a result that does not enable us to reject one hypothesis at a 5% significance level may also be quite consistent... with very different hypotheses about the population. In general, it seems wiser to focus less on tests of specific hypotheses, ... and more on best estimates and confidence intervals. For very small samples, ... the confidence intervals may prove to be wide enough to warn us that the sample results are consistent with an extremely wide range of hypotheses."

- Nevertheless, there are some environments in which one is forced, by regulations or administrative practice, to take one course of action if H_0 is rejected, and a different course if it is not rejected.

183. **Standard errors in the linear regression context.** In paragraphs 46–47, 51, it was shown how to find the best fitting straight line for predicting y from x, and how to calculate the correlation coefficient. The estimates b and a of the slope and intercept of the regression line were justified there by the least squares principle. If we spell out the model for the observations in more detail, we will be able to go on and perform hypothesis tests concerning the slope and intercept. The model is:[47]

$$y = A + Bx + \varepsilon, \tag{52}$$

where ε is a random variable whose mean is 0, whose standard deviation is σ_ε, and which has a normal distribution. (Notice this is what is called a *homoscedastic* model, meaning that σ_ε is the same for all values of x.) The estimates of B and A are b and a, given by the equations used previously. We estimate σ_ε by

$$s_\varepsilon = \sqrt{\frac{\Sigma\,(y-\hat{y})^2}{n-2}} \tag{53}$$

(the principle)

$$= \sqrt{\frac{S_{yy} - bS_{xy}}{n-2}} \tag{54}$$

(a more convenient computing formula).

[47] A lot of books use the symbols α and β for the true values of the intercept and slope, instead of A and B. But I have recently been using α and β for the probabilities of the two types of error in hypothesis testing, so I will avoid using them here.

The standard error of b is estimated as

$$\frac{s_\varepsilon}{\sqrt{S_{xx}}}, \tag{55}$$

and the standard error of a is estimated as

$$s_\varepsilon.\sqrt{\frac{1}{n}+\frac{\bar{x}^2}{S_{xx}}}. \tag{56}$$

The intercept a is what y is expected to be at $x = 0$. Sometimes, $x = 0$ has a real physical meaning. Other times, it is of no particular interest; in such a case, one might be more concerned with y at values of x close to the average value of x. At $x = \bar{x}$, the prediction $\hat{y}$ is $\bar{y}$, and the standard error is

$$\frac{s_\varepsilon}{\sqrt{n}}. \tag{57}$$

184. **Hypothesis testing of slope and intercept.** Hypothesis tests use the same idea we have met many times before: calculate

$$\frac{\text{Observed} - \text{Hypothesised}}{\text{Standard error}}.$$

This is interpreted as a t statistic with $n - 2$ degrees of freedom. Perhaps the most common hypothesis test is of H_0: $B = 0$. (If the slope could be 0, then we cannot say there is a linear relationship between x and y. Furthermore, unless we can see some particular non-linear relationship when inspecting the scatterplot, we are likely to conclude there is no relationship at all.)

185. Continuing the computations on linkage of rice markets in south China that were begun in paragraph 47,

$$\begin{aligned} s_\varepsilon &= \sqrt{\frac{0.135-(-0.0279)\times(-1.065)}{4}} \\ &= 0.162. \end{aligned}$$

And so the standard error of the slope is estimated to be

$$\frac{0.162}{\sqrt{38.235}},$$

which is 0.0262. The test statistic for testing whether the slope could be 0 is $\frac{-0.0279-0}{0.0262} = -1.06$. A t statistic with 4 degrees of freedom needs to reach 2.78 in magnitude before it is statistically-significant (when performing a two-sided test at the 0.05 level of significance). Consequently, we do not reject the null hypothesis that the slope could be 0: the data is consistent with the idea that the slope is 0.

186. **Testing whether correlation is zero.** To test whether a correlation coefficient could be 0, we refer to a table of the critical values (which, naturally, depend upon the sample size n). Such a table is printed in many textbooks of statistics. Or we can calculate

$$\frac{r}{\sqrt{1-r^2}}.\sqrt{n-2}, \tag{58}$$

and interpret this as a t statistic with $n-2$ degrees of freedom.

187. For the linkage of rice markets, r was found to be -0.47. Hence we calculate $\frac{-0.47}{\sqrt{1-(-0.47)^2}}.\sqrt{4}$, and find it is -1.06, exactly the same as we found when testing whether the slope could be 0. We therefore do not reject the null hypothesis of zero correlation.

References

The paragraph where each reference is cited is shown in square brackets.

Attenbrow, V (1992). Shell bed or shell midden. *Australian Archaeology*, No. 34, 3–21. [152]

Brown, M C, and Halaby, C N (1987). Machine politics in America, 1870–1945. *Journal of Interdisciplinary History*, 17, 587–612. [99]

Cioffi-Revilla, C (1991). The long-range analysis of war. *Journal of Interdisciplinary History*, 21, 603–629. [89]

Cowgill, G L (1975). A selection of samplers: Comments on archaeo-statistics. In J W Mueller (Editor), *Sampling in Archaeology*, pp. 258–274. Tucson: University of Arizona Press. [182]

David, F N (1955). Dicing and gaming (A note on the history of probability). *Biometrika*, 42, 1–15. (Reprinted in E S Pearson and M G Kendall (Editors), *Studies in the History of Statistics and Probability*, pp. 1–17. London: Griffin, 1970.) [68x, 81]

Donovan, J M (1991). Infanticide and the juries in France, 1825–1913. *Journal of Family History*, 16, 157–176. [179]

Floud, R, Wachter, K W, and Gregory, A (1993). Measuring historical heights — Short cuts or the long way round: A reply to Komlos. *Economic History Review*, 46, 145–154. [105x]

Freedman, D, Pisani, R, Purves, R, and Adhikari, A (1991). *Statis-*

tics. 2nd Edition. New York: Norton. [2]

Glinsman, L A, and Hayek, L C (1993). A multivariate analysis of Renaissance portrait medals: An expanded nomenclature for defining alloy composition. *Archaeometry*, 35, 49–67. [1]

Grubb, F (1992). Fatherless and friendless: Factors influencing the flow of English emigrant servants. *Journal of Economic History*, 52, 85–108. [181]

Knodel, J (1983). Seasonal variation in infant mortality: An approach with applications. *Annales de Démographie Historique*, 208–230. [168]

Komlos, J (1993). The secular trend in the biological standard of living in the United Kingdom, 1730–1860. *Economic History Review*, 46, 115–144. [105x]

Marks, R B (1991). Rice prices, food supply, and market structure in eighteenth-century south China. *Late Imperial China*, 12, 64–116. [43]

Ortloff, C R, and Kolata, A L (1993). Climate and collapse: Agroecological perspectives on the decline of the Tiwanaku state. *Journal of Archaeological Science*, 20, 195–221. [171x]

Pritchett, J B, and Freudenberger, H (1992). A peculiar sample: The selection of slaves for the New Orleans market. *Journal of Economic History*, 52, 109–127. [105x]

Richardson, L F (1944). The distribution of wars in time. *Journal of the Royal Statistical Society*, 107, 242–250. [89]

Trampuž-Orel, N, Milić, Z, Hudnik, V, and Orel, B (1991). Inductively coupled plasma-atomic emission spectroscopy analysis of metals from Late Bronze Age hoards in Slovenia. *Archaeometry*, 267–277. [25]

Index

Topics are indexed by paragraph, not by page; footnotes, Tables, and Parts are identified by fn, T, and Pt. Citations of authors should be traced with the paragraph number at the end of the entry in the list of References.

UIPS PRINTED BY THE UNIVERSITY PRINTING SERVICE

2